Convergence of Deep Learning and Artificial Intelligence in Internet of Things

This book covers advances and applications of smart technologies including the Internet of Things (IoT), artificial intelligence, and deep learning in areas such as manufacturing, production, renewable energy, and healthcare. It also covers wearable and implantable biomedical devices for healthcare monitoring, smart surveillance, and monitoring applications such as the use of an autonomous drone for disaster management and rescue operations. It will serve as an ideal reference text for senior undergraduate, graduate students, and academic researchers in the areas such as electrical engineering, electronics and communications engineering, computer engineering, and information technology.

- Covers concepts, theories, and applications of artificial intelligence and deep learning, from the perspective of the Internet of Things.
- Discusses powers predictive analysis, predictive maintenance, and automated processes for making manufacturing plants more efficient, profitable, and safe.
- Explores the importance of blockchain technology in the Internet of Things security issues.
- Discusses key deep learning concepts including trust management, identity management, security threats, access control, and privacy.
- Showcases the importance of intelligent algorithms for cloud-based Internet of Things applications.

This text emphasizes the importance of innovation and improving the profitability of manufacturing plants using smart technologies such as artificial intelligence, deep learning, and the Internet of Things. It further discusses applications of smart technologies in diverse sectors such as agriculture, smart home, production, manufacturing, transport, and healthcare.

Convergence of Deep Learning and Artificial Intelligence in Internet of Things

Edited by
Arun Kumar Rana, Ajay Rana,
Sachin Dhawan, Sharad Sharma, and
Ahmed A. Elngar

CRC Press
Taylor & Francis Group
Boca Raton London New York

CRC Press is an imprint of the
Taylor & Francis Group, an **informa** business

First edition published 2023
by CRC Press
6000 Broken Sound Parkway NW, Suite 300, Boca Raton,
FL 33487-2742

and by CRC Press
4 Park Square, Milton Park, Abingdon, Oxon, OX14 4RN

CRC Press is an imprint of Taylor & Francis Group, LLC

© 2023 selection and editorial matter, Arun Kumar Rana, Ajay Rana, Sachin Dhawan, Sharad Sharma and Ahmed A. Elngar; individual chapters, the contributors

ISBN: 978-1-032-39171-7 (hbk)
ISBN: 978-1-032-41042-5 (pbk)
ISBN: 978-1-003-35596-0 (ebk)

DOI: 10.1201/9781003355960

Typeset in Sabon
by MPS Limited, Dehradun

Contents

Editor Bios

Dr. Ajay Rana has demonstrated his intellectual, interpersonal, and managerial skills with vast teaching experience in the academics and industry with the role ranging from lecturer to professor to director to dean and senior vice president (RBEF) within a span of more than 22 years. He is an educationalist, a teacher by heart, an innovator, a strategist, and a committed philanthropist with deep organizational ethics, equality and holding hands of every individual who wishes to succeed in life as the core of his life philosophy. His areas of interest include machine learning, Internet of Things (IoT), augmented reality, software engineering and soft computing. He has 60-plus patents under his name in the field of IoT, networks, and sensors. He has published more than 271 research papers in reputed journals such as ACM, Springer, Elsevier, Taylor and Francis and others, and International and National Conferences, co-authored 8 books and co-edited 36 conference proceedings. Prof. Rana is the chairman of AUN Research Labs, executive committee member of IEEE Uttar Pradesh Section, senior member of IEEE, and life member of the Computer Society of India and ISTE. He is also a member of the editorial board and review committee of several journals. With the intent to provide a robust learning environment to his students, he has visited many top universities and colleges, which have a legacy of more than 1,200 years of producing leaders on the global front in entire Europe, UAE, the UK, and Asia.

Dr. Sharad Sharma is presently working as a professor in Electronics and Communication Engineering Department in Maharishi Markandeshwar (Deemed to be University), Mullana, India. He has a vast experience of more than 20 years of teaching and administrative work. He has received his B.Tech degree in Electronics Engineering from Nagpur University, Nagpur, India; M.Tech in Electronics and Communication Engineering from Thapar Institute of Engineering and Technology, Patiala, India; and PhD from National Institute of Technology, Kurukshetra, India. He has a teaching experience of more than 20 years. He has conducted many workshops on Soft Computing and its applications in engineering, wireless networks, simulators, etc. He has a keen interest in teaching and implementing the latest techniques related to wireless and mobile communications. He has opened up a student chapter of IEEE as a branch counsellor. His research interests are routing protocol design, performance evaluation, and optimization for wireless mesh networks using nature-inspired computing, Internet of Things and space communication, etc. He is a member of various professional bodies.

Dr. Ahmed A. Elngar is an assistant professor of Computer Science at the Faculty of Computers and Artificial Intelligence, Beni-Suef University, Egypt. Dr. Elngar is the founder and head of Scientific Innovation Research Group (SIRG). Dr. Elngar is the director of the Technological and Informatics Studies Center (TISC), Faculty of Computers and Artificial Intelligence, Beni-Suef University. Dr. Elngar has more than 55 scientific research papers published in prestigious international journals and over 25 books covering such diverse topics as data mining, intelligent systems, social networks, and smart environment. Dr. Elngar is a collaborative researcher. He is a member of the Egyptian Mathematical Society (EMS) and the International Rough Set Society (IRSS). His other research areas include the Internet of Things (IoT), network security, intrusion detection, machine learning, data mining, artificial intelligence, big data, authentication, cryptology, healthcare systems, and automation systems. He is an editor and reviewer of many international journals around the world. Dr. Elngar won several awards including the "Young Researcher in Computer Science Engineering," from Global Outreach Education Summit and Awards 2019, on 31 January 2019 (Thursday) in Delhi, India. Also, he awards the "Best Young Researcher Award (Male) (Below 40 years)" and Global Education and Corporate Leadership Awards (GECL-2018).

Dr. Prof. Arun Kumar Rana has received his B.Tech degree from Kurukshetra University; M.Tech degree from Maharishi Markandeshwar (Deemed to be University), Mullana, India; and PhD from Maharishi Markandeshwar (Deemed to be University), Mullana, India (Final viva awaited). His area of interest includes image processing, wireless sensor network, Internet of Things, artificial intelligence, and machine learning and embedded systems. Mr. Rana has worked with Panipat Institute of Engineering and Technology, Samalkha, Haryana, India. Currently he is working as an assistant professor with Galgotias College of Engineering & Technology, Greater Noida, Uttar Pradesh, India. He has published around 70 SCI/ESCI/Scopus papers in national and international journals, and also at conferences. He has also published six books with national and international publishers and many times members of SCI/Scopus indexed international conferences/symposiums. Also, he has attended nine workshops and nine FDPs. He has guided six M.Tech candidates and one candidate is in process. He serves as a reviewer for several journals and international conferences. He is also a member of the Asia Society of Research. He also published six national and international patents. He has received many international awards from various organizations. He was listed in the world scientist ranking 2021 and 2022. He is the guest editor for Special Issue "Routing and Protocols for Energy Efficient Communication" energy, MDPI, SCI, IF-3004.

Dr. Prof. Sachin Dhawan has received his B.Tech degree from Kurukshetra University, Kurukshetra, in 2008; M.Tech degree in Electronics and Communication Engineering from the University Institute of Engineering and Technology under Kurukshetra University, Kurukshetra, in 2011; and PhD degree in Electronics and Communication Engineering from AIACT&R under Guru Gobind Singh Indraprastha University, Dwarka, New Delhi (Final viva awaited). He was an assistant professor at Geeta Institute of Management and Technology, Kanipla, Kurukshetra, from July 2011 to July 2012. Currently, he is working with Panipat Institute of Engineering and Technology, Samalkha, since July 2012 as an assistant professor. He has authored over 40 research papers in various international journals and conferences. He has also published his papers in IEEE/Springer/conferences and various Scopus/SCI journals. His current research interests include signal and image processing, Internet of Things, artificial intelligence, and machine learning. He has also published two books with national and international publishers and many times members of SCI Scopus indexed international conferences/symposiums. He is also an associate member of All India Institutions of Engineers. He serves as a reviewer for several journals and international conferences. He is also a member of the Asia Society of Research. He has also published two patents.

Contributors

Amrita Singh
College of Information
   Technology and Engineering
Nepal

Anita Dahiya
Panipat Institute of Engineering
   and Technology
Samalkha, Haryana, India

Anuradha D. Thakare
Pimpri Chinchwad College
   of Engineering
Pune, India

Arun Kumar Rana
Department of ECE
Galgotias College of Engineering &
   Technology,
Greater Noida, India

Asutosh Goswami
Department of Geography
Lovely Professional University
Phagwara, Punjab, India

A. Sugunapriya
Sri Ramanujar Engineering
   College
Kolapakkam, Tamil Nadu, India

Dr. Anuradha
Department of Computer Engineering
Pimpri Chinchwad College of
   Engineering
Pune, India

C. Pinto
Goa University
Taleigao Plateau, Goa, India

Dr. Cheshta Jain Khare
EED SGSITS
Indore, Madhya Pradesh, India

F. Antony Xavier Bronson
Department of Computer Science
Dr. M.G.R. Educational and
   Research Institute
Chennai, Tamil Nadu, India

G. Boopathi Raja
Department of ECE
Velalar College of Engineering
   and Technology
Erode, Tamil Nadu, India

G. Senthil Velan
Department of Computer Science
Dr. M.G.R. Educational and
   Research Institute
Chennai, Tamil Nadu, India

G. M. Naik
Goa University
Taleigao Plateau, Goa, India

Dr. H. K. Verma
EED SGSITS
Indore, Madhya Pradesh, India

Ishfaq Majid
School of Education
Central University of Gujarat
Gujrat, India

Jivan Parab
Goa University
Taleigao Plateau, Goa, India

Kashif Nisar
Universiti Malaysia Sabah
Jalan UMS
Kota Kinabalu, Sabah, Malaysia

K. Thippeswamy
Visvesvaraya Technological
   University
Mysuru, Karnataka, India

Manju Rani
Department of ASH
Panipat Institute of Engineering
   and Technology
Samalkha, Haryana, India

Marlon
School of Physical and
   Applied Science
Goa University
Goa, India

M. Lanjewar
Goa University
Taleigao Plateau, Goa, India

M. Sequeira
Goa University
Taleigao Plateau, Goa, India

Monish Gupta
University Institute of Engineering
   and Technology
Kurukshetra University
Kurukshetra, Haryana, India

Navneet Kaur
Department of ECE
Chandigarh Group of Colleges
Mohali, Punjab, India

Naveen Kumar
Department of ASH
Panipat Institute of Engineering
   and Technology
Samalkha, Haryana, India

Nidhi Chahal
Chandigarh Engineering College
Chandigarh, India

P. Solainayagi
Department of Computer Science
Aarupadai Veedu Institute
   of Technology
Paiyanoor, Chennai

Priya Trivedi
Compfeeders Aisect College
   of Professional Studies
Indore DAVV
Madhya Pradesh, India

Paras Chawla
iNurture Education Solutions Pvt Ltd
Niton Compound, Palace Road
Banglore, India

Pranjali G. Gulhane
Pimpri Chinchwad College
   of Engineering
Pune, India

P. Karthika
Kalasalingam Academy of
   Research and Education
Tamil Nadu, India

Dr. Renu Bala
JCDM – Jan Nayak Chaudhary
   Devi Lal Memorial College
Sirsa, Haryana, India

Rishabh Jha
College of Information Technology
and Engineering
Nepal

R. Ganesh Babu
SRM TRP Engineering College
Tiruchirappalli, Irungalur
Tamil Nadu, India

Rupa Sanyal
Department of Botany
Bhairab Ganguly College
Kolkata, West Bengal, India

Saravanan Elumalai
Department of Computer Science
Dr. M.G.R. Educational and
Research Institute
Chennai, Tamil Nadu, India

Sachin Dhawan
Department of ECE
Panipat Institute of Engineering
and Technology
Samalkha, Haryana, India

Sandeep Kajal
Kyung Hee University
South Korea

Sumit Kumar Rana
Department of CSE
Panipat Institute of Engineering
and Technology
Samalkha, Haryana, India

Suhel Sen
Department of Geography
Bhairab Ganguly College
Kolkata, West Bengal, India

S. Markkandan
SRM TRP Engineering College
Tiruchirappalli, Irungalur
Tamil Nadu, India

V. Sai Shanmuga Raja
Department of Computer Science
Shanmuganathan Engineering
College
Arasampatti, Pudukottai District
Tamil Nadu, India

Vishal Gupta
University Institute of Engineering
and Technology
Kurukshetra University
Kurukshetra, India

Dr. Vikas Khare
NMIMS
Indore, Madhya Pradesh, India

Dr. Y. Vijaya Lakshmi
School of Education
Central University of Gujarat
Gujrat, India

Chapter 1

Deep Learning Algorithms for Real-Time Healthcare Monitoring Systems

G. Boopathi Raja

CONTENTS

DOI: 10.1201/9781003355960-1

1.1 INTRODUCTION

One of the most important aspects of human life is health. Health is a benefit when it comes to avoiding disease and getting a better physiological state. Every civilization is paying increasing attention to and implementing technology in the area of health and healthcare [1–3]. COVID-19 has recently had a significant impact on every country's economy, necessitating the introduction of smart healthcare systems. People are remotely controlled in smart healthcare systems to avoid the spread of sickness and to offer timely and cost-effective treatment. In this scenario, the optimum approach appears to be merging internet of things (IoT)-enabled healthcare systems with machine intelligence [4].

Because of progressions in sensors, computation, spectrum usage, and machine intelligence, IoT and machine learning (ML)-based solutions are effective. Advances in microelectronics have enabled the development of compact and affordable medical sensing devices, providing healthcare support [5]. As a consequence, healthcare providers categorize these therapies into two groups: symptomatic and preventative. People nowadays devote a lot of attention to illness prevention and early detection, as well as the best treatment options for numerous chronic conditions. As a result, national healthcare monitoring devices are becoming an unavoidable trend these days.

In the field of telemedicine, IoT-connected devices and deep learning (DL) algorithms for continuously monitoring patients for proper diagnosis have recently sparked a lot of interest. Real-time healthcare systems that are energy-efficient and adaptive are in high demand to regulate people's vital signs [6]. Traditional mobile communication for healthcare systems, on the other hand, is subject to radiation and has a greater cost. Real-time health monitoring systems, on the other hand, are radiation-free, have multiple communication channels, and can monitor a wide range of conditions. Data connection and transmission may be accomplished in the house by connecting to a wireless router. When people are out in the open, however, the handheld healthcare systems are attached via a wireless medium to enable data information exchange and transmission. In this case, ML algorithms are being used to determine whether the people are indoors or outdoors.

ML and DL algorithms are rapidly being applied in IoT-enabled healthcare systems [7]. Identical, lossy, and noisy data is stored and processed in far-flung cloud data centres, squandering resources, and causing catastrophic health-related events [8]. These algorithms ensure that relevant characteristics are extracted from large volumes of raw data and that the optimal choice is made for time-critical and delay-sensitive activities in sports informatics [9].

Standard techniques are necessary to analyze and extract meaningful information from the vast amount of diverse data collected in sports informatics. Using traditional learning techniques, the deep neural network (DNN) can immediately produce meaningful features from an unstructured dataset without the need for human intervention or engineering [10–12].

This chapter presents the various types of DL algorithms used in practice with suitable examples. Section 1.2 introduces the concept of DL approaches along with functionalities and application areas. Section 1.3 describes the DL algorithm, the definition of the basic neural network, and its implementation. Different types of DL algorithms are introduced and briefly explained in Section 1.4. The hardware requirements for the implementation of these algorithms are discussed in Section 1.5. Finally, the chapter is concluded in Section 1.6.

1.2 DEEP LEARNING

A three-layer neural network is used in DL, which is an ML approach. These neural networks attempt to emulate human brain activity by permitting them to learn from enormous amounts of data, yet they frequently fall short of the capabilities of the human brain. A single-layer neural network can only make informed estimates at this point, but adding additional hidden layers can help optimize and improve performance.

ML and DL models may do supervised learning, unsupervised learning, and reinforcement learning. Supervised learning classifications or predictions using labelled datasets; this requires human intervention to appropriately categorize incoming data. Unsupervised learning, on the other hand, does not need labelled datasets; instead, it looks for patterns in data and organizes it according to whatever criteria it chooses. Reinforcement learning is a type of learning in which a system increases its accuracy for doing a particular action in a specific environment depending on feedback to maximize the reward.

1.2.1 How Does DL Function?

DL neural networks, also known as artificial neural networks, strive to emulate the human brain by integrating data inputs, bias, and weights. These parts work together to recognize, categorize, and classify data pieces accurately.

A neural network can provide forecasts and compensate for errors via forward and reverse propagation. As time passes, the algorithm increases in accuracy. DL methods are extremely difficult to implement, and various sorts of neural networks are used to solve specific problems or datasets. As an illustration,

- Convolutional neural networks (CNNs), which are commonly employed in learning algorithms and image analysis applications. It may detect and recognize objects by recognising features and patterns inside an image.

- Recurrent neural networks (RNNs) are frequently employed in natural language and voice recognition applications, despite the fact that they need sequence or time series data.

1.2.2 Applications of DL

In the real world, DL applications are prevalent, however, they are frequently so well integrated into products and services that customers are unaware of the substantial data analysis that occurs behind scenes. The following are a few examples.

1.2.2.1 Law Enforcement

DL systems can learn from transactional data to detect possible fraudulent or illegal behaviour. Computer vision, images, speech recognition, documents, and certain other DL implementations may improve the performance based on investigative analysis. This is possible by extracting patterns and substantiation from sound and video recordings, allowing law enforcement to analyse massive quantities of data more rapidly and correctly.

1.2.2.2 Financial Services

Predictive analytics is used by financial organisations to enhance algorithmic stock trading, evaluate company risks for loan approvals, identify fraud, and assist customers with credit and financial portfolio management.

1.2.2.3 Healthcare

With the digitalization of medical data and photographs, DL capabilities have substantially improved the healthcare business. Medical imaging professionals and radiologists profit from image recognition software because it allows them to study and analyse more images in less time.

1.3 DL ALGORITHMS

DL is a type of ML and it mimics the human brain. DL has risen in popularity in numerical computation, with its methodologies being used to solve complex problems across a variety of industries. DL networks learnt from the environment or situations are utilized to train machines. Figure 1.1 shows the basic concept involved in neural networks.

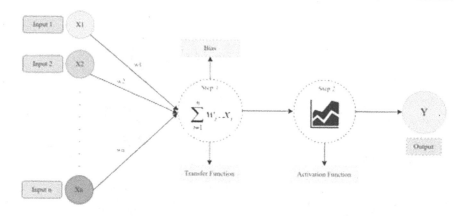

Figure 1.1 Basic concept of neural network.

Figure 1.2 shows the different types of smart wearable devices placed in various parts of the human body in order to continuously monitor the physiological parameters of that person. Based on the parameters received, decision may be taken with the help of a DNN.

1.4 TYPES OF ALGORITHMS USED IN DL

1. CNNs
2. RNNs
3. Long short-term memory networks (LSTMs)
4. Multi-layer perceptrons
5. Radial basis function networks (RBFNs)
6. Self-organizing maps (SOMs)
7. Generative adversarial networks (GANs)
8. Autoencoders (AEs)
9. Restricted Boltzmann Machines(RBMs)
10. Deep belief networks (DBNs)

1.4.1 Convolutional Neural Networks

CNNs are multi-layer neural networks used largely for image processing and object recognition. It was able to recognize characters such as ZIP codes and numbers. A CNN is a DL system that can take an input image and assign biases to various aspects/objects in the image, as well as discern between them. The quantity of pre-processing needed by a ConvNet is much less than that required by other classification techniques. Figure 1.3 shows the architecture of CNN [13].

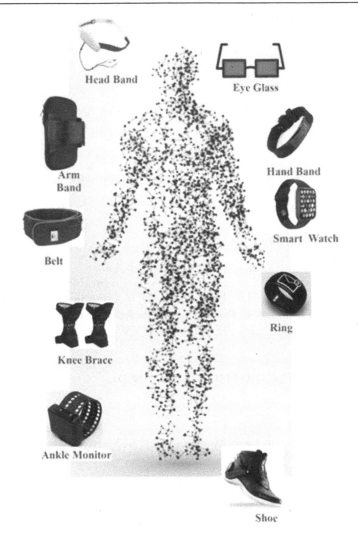

Figure 1.2 Smart wearable devices.

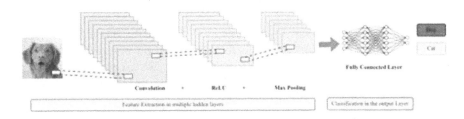

Figure 1.3 Convolution neural network.

The architecture provides better fitting to the image dataset due to a smaller number of input parameters and the recyclability of weights. Multiple layers in CNNs process and retrieve data features.

1.4.1.1 Convolution Layer

Convolution operation is used to retrieve high-level characteristics such as edges from an input image. In a ConvNet, the amount of convolutional layers does not have to be limited to one. The first ConvLayer is frequently responsible for collecting low-level data such as edges, colour, gradient direction, and so on.

1.4.1.2 Pooling Layer

Like the convolutional layer, the pooling layer is important for reducing the spatial size of the convolved feature. The computer power required to analyze huge volumes of data is decreased by dimensionality reduction. It also helps maintain the model's training process going smoothly by collecting rotational and positional invariant dominant features.

1.4.1.3 Fully Connected Layer

Using a fully connected layer to acquire non-linear permutations of high-level information carried by the convolutional layer's output is a (usually) low-cost method. The fully connected layer is developing a possibly non-linear function in this region.

1.4.2 Long Short-Term Memory Networks

LSTMs, a type of RNN, may be used to learn and recall long-term dependencies. The default mode of operation is to recall earlier information over a long period of time. Figure 1.4 describes the architecture of LSTM.

This model is divided into cells, each of which has numerous operations. In the LSTM, operation gates change an internal process variable that is sent through one cell to the next.

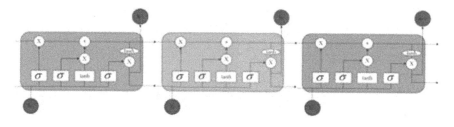

Figure 1.4 Long short-term memory networks.

Data are tracked using LSTMs throughout time. Four interrelated information of the connections in a unique form is provided by LSTMs, which give a chain-like structure. LSTMs are widely employed in voice recognition, music creation, and pharmaceutical research when it comes to time forecasts [14].

1.4.2.1 Working of LSTMs

- First, they forget about non-essential aspects of the prior condition.
- They then modify the cell state values selectively.
- Finally, certain aspects of the cell state's output.

1.4.2.1.1 Forget Gate

$$f_t = \sigma(W_f. [h_{t-1}, \ x_t] + b_f)$$

It is a sigmoid layer that concatenates the output at time $t - 1$ with the input sequence at time t into a single tensor before applying a linear transformation and a sigmoid. The outcome of this gate is somewhere between 0 and 1 due to the sigmoid.

1.4.2.1.2 Input Gate

$$i_t = \sigma(W_i. [h_{t-1}, \ x_t] + b_i)$$

The old output and the new input are passed through another sigmoid layer via the input gate. A value between 0 and 1 is returned by this gate. The outcome of the candidate layer is multiplied by the input signal gate.

$$C_t = tanh(W_C. [h_{t-1}, \ x_t] + b_C)$$

This layer adds a candidate vector to the internal state by applying a hyperbolic tangent on the combination of input and prior output.

This rule is used to update the internal state:

$$C_t = f_t * C_{t-1} + i_t * C_t$$

The forget gate multiplies the prior state, which is then contributed to the percentage of the potential applicant that the output gate allows.

1.4.2.1.3 Output Gate

$$O_t = \sigma\,(W_O.\,[h_{t-1},\ x_t] + b_O)$$

$$h_t = O_t * tanh\,C_t$$

This gate functions similarly to the others in that it regulates that most of the internal state is transferred to the response.

1.4.3 Recurrent Neural Networks

Recurrence is a mathematical concept that describes the sort of reliance of the present value (event or word) also on prior event(s) and is stated using recurrent equations.

$$y_k = f\,(y_{k-1},\, x_k)$$

A recurrent neural network is made up of several copies with the same node, each of which sends a message to the next node in the chain. Each unit comprises three sets of weights: one with the inputs $x(t)$, one for the previous time step's outputs $y(t-1)$, and one for the current time step's output $y(t)$. Figure 1.5 represents the basic architecture of RNNs [14].

The LSTM's output is used as an input to the current mode, and it has internal memory that allows it to recall previous inputs. RNNs are commonly used for handwriting recognition, natural language processing, time

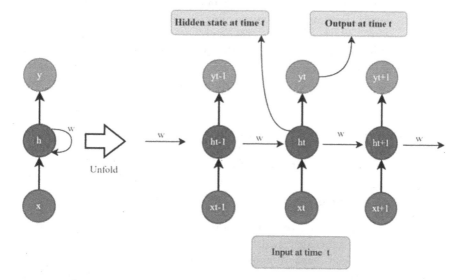

Figure 1.5 Recurrent neural networks.

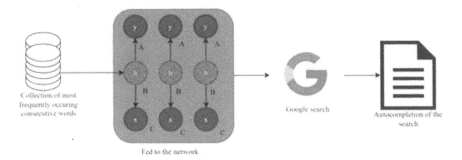

Figure 1.6 Working of recurrent neural networks.

series analysis, image captioning, and machine translation. Figure 1.6 shows the example of implementation of RNN.

1.4.4 Generative Adversarial Networks

GANs are generative models developed in 2014 by Goodfellow et al. Two differentiable functions, modelled by neural networks, are engaged in a game in a GAN setup. It can generate new data instances that are similar to the training data [15]. GAN consists of two parts: a generator that trains to produce false data and a discriminator that further acquires from the data. In this framework, the two participants (the generator and the discriminator) play separate roles.

The generator attempts to generate data from a probability distribution. The discriminator takes on the role of a judge. It has the authority to choose whether the input comes out from the generator or the genuine training set. In simple words, generator and discriminator are described as follows:

- The generator tries to increase the chances of the discriminator mistaking its inputs for real.
- The discriminator, on the other hand, directs the generator to generate more realistic pictures.

GANs help in the production of cartoon characters and realistic images, as well as the depiction of 3D objects and photographs of human faces. Figure 1.7 shows the architecture of GAN.

1.4.4.1 How Do GANs Work?

The discriminator learns to tell the difference between the bogus data generated by the generator and the genuine sample data. The generator generates fraudulent data during early training, and the discriminator soon learns to recognize it as such. To update the model, the GAN delivers the findings towards the generator and discriminator.

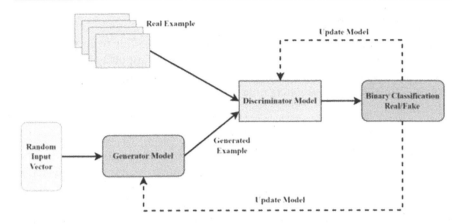

Figure 1.7 Generative adversarial networks.

1.4.5 Radial Basis Function Networks

The RBF is a mathematical function that an artificial neural network that likes to use radial basis functions as operational functions is known as a neural network. The kernel function of the inputs and the neuron parameters are combined in the output obtained by this function. It is useful for a variety of tasks, including time series prediction, classification, function approximation, and system control. As a result, it is tremendously beneficial to AI startups. Figure 1.8 shows the architecture of RBFN.

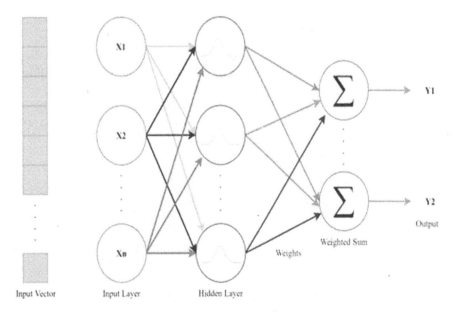

Figure 1.8 Architecture of RBFNs.

1.4.5.1 How Do RBFNs Work?

RBFNs use the resemblance of the inputs to samples out from training set to classify it. An input vector supplies data to the input layer of RBFNs. They have an RBF neuron layer. The output layer comprises one node per group, and the function determines the weighted total of the inputs.

1.4.6 Multi-Layer Perceptrons

Researchers can start by layering linear perceptrons on top of each other to answer increasingly complicated and non-linear issues. Multi-layer perceptrons are a network of stacked perceptrons or deep artificial neural networks. Due to the existence of numerous layers, it is described as "deep." And "network" because of the layered arrangement's collaboration towards a shared purpose [16]. Figure 1.9 shows the architecture of MLP with a simple demonstration.

An MLP is made up of an input layer, an output layer, and a specified number of hidden layers in between. Depending on the application, the number of hidden layers varies. Also, much as in a single or linear perceptron, each node in the input and hidden layers has a weight associated with it.

MLPs are a sort of feed-forward neural network. It is made up of multiple layers of perceptrons with non-linear activation functions. It consists of two layers that are entirely interconnected: an input layer and an output layer.

1.4.6.1 Working of MLPs

The data are fed into the network's input layer using MLPs. The signal flows in one way because the layers of neurons are connected in a graph. All activation functions are sigmoid functions, *tanh*, and ReLUs.

An MLP is shown below as an example. To categorize photos of cats and dogs, the network identifies the weight values and performs appropriate activation functions.

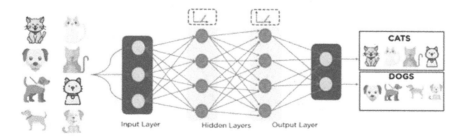

Figure 1.9 Illustration of MLP.

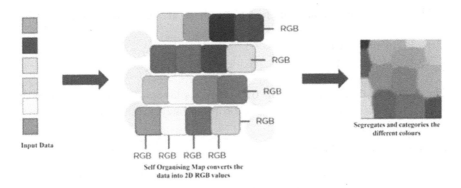

Figure 1.10 Self-organizing maps.

1.4.7 Self-Organizing Maps

SOMs are a sort of unsupervised learning neural network. It is an efficient algorithm used to recognize the features. They are used to construct a low-dimensional training sample space. It uses self-organizing artificial neural networks for the reduction of dimensionality of data, allowing for better data visualisation. As a result, they might be used to cut down on the number of dimensions. SOMs differ from conventional artificial neural networks in that they use competitive learning instead of error-correlated learning, which relies on backpropagation and gradient descent [17]. Figure 1.10 shows the architecture of SOMs.

1.4.7.1 Working of SOMs

SOMs pick an array at random out from the learning phase to establish weights for each node. The best matching unit (BMU) is the winning node. The BMU's neighbourhood is discovered through SOMs, and the number of neighbours decreases with time. The sample vector is given a winning weight using SOMs. The weight of a node varies as it gets closer to a BMU.

The more away a neighbour is from the BMU, the less it learns from it. For *N* iterations, SOMs repeat step two. Figure 1.10 shows the colour-coded representation of an input array. These data may be sent into an SOM, which converts it to RGB values in two dimensions. Finally, it categorizes and distinguishes the many hues of colour.

1.4.8 Deep Belief Networks

DBNs are generative models that contain many layers of stochastic, latent variables. Hidden units, also known as latent variables, have binary values [18].

The pre-train phase and the fine-tune phase are the two phases of DBNs.

- Pre-train Phase
- Fine-tune Phase

The pre-train phase is made up of numerous layers of RBNs, whereas the fine-tune phase is made up of a feed-forward neural network.

1.4.8.1 Working of DBNs

Greedy learning approaches are used to train DBNs. To learn the top-down, generative weights, the greedy learning approach employs a layer-by-layer process. On the two largest buried layers, DBNs perform Gibbs sampling steps. The RBM described by the highest two hidden layers is sampled in this step. DBNs learn that a single bottom-up pass may infer the quantities of the dependent variable in each layer. The following illustration is based on the architecture of DBN, as shown in Figure 1.11.

1.4.9 Restricted Boltzmann Machines

RBM is a type of neural network that uses part of the energy-based model. It is a generative deep ML technique that's probabilistic, unsupervised, and probabilistic. The goal of RBM is to discover the joint probability distribution that maximizes the logarithm of the likelihood function. RBM is a non-directed algorithm with only two layers: input and hidden. Every visible node is linked to every hidden node. RBM is a symmetrical bipartite graph because it contains two layers: a visible layer or input layer and a hidden layer [18].

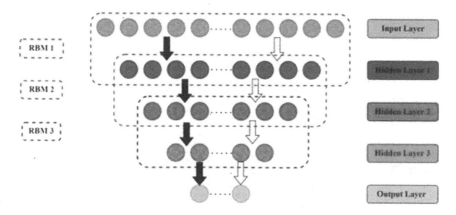

Figure 1.11 Deep belief networks.

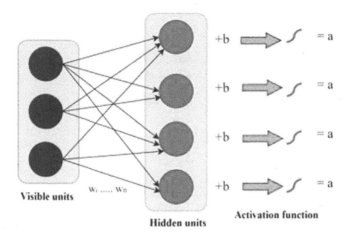

+b = a
+b = a
+b = a
+b = a

Visible units
W₁ ..... Wn
Hidden units
Activation function

Figure 1.12 Restricted Boltzmann machines.

RBMs are made up of two layers:

- Hidden layers
- Visible layers

The RBM assesses the output quality by analyzing the restoration to the original data at the visible layer. Figure 1.12 is a diagram showing how RBMs work.

1.4.10 Autoencoders

AEs consist of the same input and output as a feed-forward neural network. AEs are neural networks that duplicate their sources into its responses. The objective of this network is to convert the data into a latent space description. Finally, the result may be reconstructed/restored from that representation [19]. Basically, there are only two parts available in this network:

- The encoder is the network component that reduces the amount of the input into a latent space representation. The function used for encoding operation is given by $h = f(x)$.
- Decoder: The goal of this section is to rebuild the input using the latent space representation. The function used for decoding operation is given by $r = g(h)$.

1.4.10.1 How Do AEs Work?

AEs are devices that take in data and convert it into a new format. After that, they try to duplicate the data input as closely as possible. When a

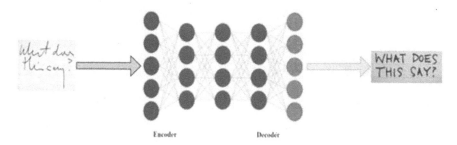

Figure 1.13 Autoencoders.

digit's image is not clear, it is then sent to an AE neural network to be processed. Before compressing the data into a smaller package, AEs initially encode the picture. Finally, the image is decoded by the AE, yielding the restored image. Figure 1.13 describes the operation of AEs.

1.5 HARDWARE REQUIREMENTS OF DL

DL needs a significant amount of processing power. High-performance graphics processing units (GPUs) are suitable because they may manage a huge number of operations in a big number of cores while still having enough storage. Managing a large number of GPUs on-premises, but on the other hand, might put a burden on current resources and be very expensive to scale.

1.6 CONCLUSION

DL algorithms have been much more popular in a variety of industries during the last five years. After authentication, information on physiological parameters from wearable sensor devices is immediately saved to an actual medical health monitoring system. Furthermore, the harmful influences are avoided by using a well-designed router wireless sensor network that operates at low power. The application of the DNN model provided encouraging results, although the DNN model has a number of drawbacks. To begin with, the system is sensitive to the level of neurons in each hidden layer, which means that a few neurons might produce under-fitting issues.

REFERENCES

[1] Wu, X., Liu, C., Wang, L., et al. (2021). Internet of things-enabled real-time health monitoring system using deep learning. *Neural Computing and Application.* 10.1007/s00521-021-06440-6.

[2] Raja, G. B. (2021). Impact of internet of things, artificial intelligence, and blockchain technology in Industry 4.0. In Kumar, R., Wang, Y., Poongodi, T., Imoize, A. L. (eds.) *Internet of Things, Artificial Intelligence and Blockchain Technology*. Cham: Springer. 10.1007/978-3-030-74150-1_8.

[3] Boopathi Raja, G. (2021). *Impact of IIOT in Future Industries: Opportunities and Challenges, Internet of things* (1st ed.). USA: CRC Press.

[4] Liu, X., He, P., Chen, W., & Gao, J. (2019) Multi-task deep neural networks for natural language understanding. *arXivdoi*. 10.18653/v1/p19-1441.

[5] Sun, J., Khan, F., Li, J., Alshehri, M. D., Alturki, R., & Wedyan, M. (2021). Mutual authentication scheme for ensuring a secure device-to server communication in the internet of medical things. *IEEE Internet Things Journal*, 10.1109/JIOT.2021.3078702.

[6] Zhang, Q., Yang, L. T., Chen, Z., & Li, P. (2017). An improved deep computation model based on canonical polyadic decomposition. *IEEE Transaction on Systems, Man, and Cybernetics: Systems*, 48(10), 1657–1666.

[7] Jan, M. A., Khan, F., Khan, R., Watters, P., Alazab, M., & Rehman, A. U. (2021). A lightweight mutual authentication approach for intelligent wearable devices in health-CPS. *IEEE Transactions on Industrial Informatics*, 17(8), 5829–5839.

[8] Jan, S. R., Khan, R., Khan, F., & Jan, M. A. (2021). Marginal and average weight-enabled data aggregation mechanism for the resource constrained networks. *Computer Commununication Journal*, 174, 101–108.

[9] Jan, M. A., Khan, F., Mastorakis, S., Adil, M., Akbar, A., & Stergiou, N. (2021). LightIoT: Lightweight and secure communication for energy-efficient IoT in health informatics. *IEEE Transactions on Green Communication and Networking*. 10.1109/TGCN.2021.3077318.

[10] Boopathi Raja, G. (2021). *Fingerprint based Smart Medical Emergency First Aid Kit using IOT, Electronic Devices, Circuits, and Systems for Biomedical Applications* (1st edition). USA: Academic Press. 10.1016/B978-0-323-851 72-5.00015-0.

[11] Boopathi Raja, G. (2021). *Appliance Control System for Physically Challenged and Elderly Persons through Hand Gesture-Based Sign Language, Biomedical Signal Processing for Healthcare Applications* (1st ed.). USA: CRC.

[12] Boopathi Raja, G. (2022). Early detection of breast cancer using efficient image processing algorithms and prediagnostic techniques: A detailed approach. *Cognitive Systems and Signal Processing in Image Processing*, 223–251.

[13] Alzubaidi, L., Zhang, J., Humaidi, A. J., et al. (2021). Review of deep learning: Concepts, CNN architectures, challenges, applications, future directions. *Journal of Big Data*, 8, 53. 10.1186/s40537-021-00444-8.

[14] Bouktif, S., Fiaz, A., Ouni, A., & Serhani, M. A. (2020). Multi-Sequence LSTM-RNN deep learning and metaheuristics for electric load forecasting. *Energies*, 13(2), 391. 10.3390/en13020391.

[15] Song, M., Zhang, J., Chen, H., & Li, T. (2018). Towards efficient microarchitectural design for accelerating unsupervised GAN-based deep learning. *2018 IEEE International Symposium on High Performance Computer Architecture (HPCA)*, pp. 66–77. 10.1109/HPCA.2018.00016.

[16] Ben Driss, S., Soua, M., Kachouri, R., & Akil, M. (1 May 2017). A comparison study between MLP and convolutional neural network models for character recognition. *Proceedings of SPIE 10223, Real-Time Image and Video Processing, 2017,* 1022306. 10.1117/12.2262589.

[17] Ali, M. N., Sarowar, M. G., Rahman, M. L., Chaki, J., Dey, N., & Tavares, J. M. (2019). Adam deep learning with SOM for human sentiment classification. *International Journal of Ambient Computing and Intelligence (IJACI), 10*(3), 92–116. 10.4018/IJACI.2019070106.

[18] Zhang, X. & Chen, J. (2017). Deep learning based intelligent intrusion detection. *2017 IEEE 9th International Conference on Communication Software and Networks (ICCSN),* pp. 1133–1137. 10.1109/ICCSN.2017. 8230287.

[19] Raghavan, P. & Gayar, N. E. (2019). Fraud detection using machine learning and deep learning. *2019 International Conference on Computational Intelligence and Knowledge Economy (ICCIKE),* pp. 334–339. 10.1109/ ICCIKE47802.2019.9004231.

Chapter 2

Neural Network–Based Efficient Hybrid Control Scheme for the Tracking Control of Autonomous Underwater Vehicles

Manju Rani, Naveen Kumar, and Anita Dahiya

CONTENTS

2.1 INTRODUCTION: BACKGROUND

In recent decades, the studies as well developments of autonomous underwater vehicles (AUVs) have gained much attention in the execution of many undersea applications such as geological sampling, underwater pipelines inspection, deep sea archaeology, oceanographic mapping, underwater surveillance, ocean floor survey, and geological sampling [1–6]. Without putting living beings in danger, AUVs and remotely operated vehicles (ROVs) work effectively when contrasted with human-involved vehicles (HOVs). Path following control difficulty is a good deal easy and worried with the control legal guidelines that make the vehicle to comply with a geometrical manner [7–9]. But, due to the existence of uncertainties and external disturbances, the development of the control technique for the trajectory tracking has become a difficult task and has gained a wide range of attention among researchers [10]. Different methodologies were accounted in the literature for the proper movement control of AUVs including model-reliant control, backstepping control, sliding mode control, and adaptive/robust control [11–20]. A little bit of them is studied here. Proportional-Integral-Derivative (PID) regulators are consistently utilized for the control of AUVs because of the general simplicity of execution. The model-reliant adaptive control techniques were effectively produced for AUVs in [11,12]. For the tracking of the completely impelled submerged vehicles, a model-reliant control scheme was planned by the authors in [13].

DOI: 10.1201/9781003355960-2

From the writing overview, it has been seen that the model-based controllers encompass accurate information of the vehicle and can't deliver enormous stage trajectory tracking manage accuracy. Besides, in reasonable applications, the specific unique data can't be accomplished. To improve the strength against the uncertainties, robust/adaptive control plans were planned and some other control strategies were proposed with no need for the specific unique model of the AUVs. In [14], the control issue of independent submerged vehicles was settled by a high-order adaptive sliding mode regulator. In the introduced scheme, through real-time examinations, the vigour of the introduced control strategy was affirmed. A robust control technique was considered for the time-changing development control of numerous underactuated AUVs [15]. The uniform extreme boundedness of all signs was ensured by the planned control law.

However, from the literature review, it's obvious that these controllers require the reckoning of the regression matrix and this framework similarly relies on the managed framework. For the trajectory tracking control issue of underactuated vehicles, an adaptive and novel sliding mode control technique was introduced in [16]. In this plan, to improve the power of the control method, sliding mode control and the backstepping technique were coupled. In [17], a sliding mode control-based trajectory tracking issue was settled. In this, a disturbance observer was additionally planned for the assessment of the time-changing obliged unsettling disturbances. A terminal sliding mode control law was considered in [18] for lateral movement control of underactuated AUVs. By coupling integral terminal and quick integral sliding mode controls, a direction following plan was introduced in [19] for unmanned underwater vehicles (UUVs). After this, an effective adaptive quick non-singular essential terminal sliding mode manage law was created in [20] for the direction following control of AUVs and ensured a quicker convergence rate. Despite the fact that an asymptotic following exhibition was effectively given by the sliding mode control for the unsure non-straight frameworks, but its resultant controller is irregular [21–23]. An incorporated sliding mode and backstepping control strategy were considered for the following control of UUVs in [22,23]. From the literature audit, it is obvious that for the unsure non-linear dynamics of AUVs, diverse analysts have focused on the utilization of the intelligent equipment control and their accumulation with the opposite control strategies for following of AUVs [24–30]. An objective following control was addressed in three-dimensional space with the pre-arranged execution of the underactuated AUVs [24]. Depending upon the repetitive neural organization, the authors in [25] introduced an online straight control technique. In this law, the network started from randomly primary conditions with no need of past training. In [26], an NN adaptive technique was introduced for the control of AUVs. The adequacy of the planned control technique was assessed through simulation considerations. In [27], for the sophisticated position control, sliding mode NN control law was introduced for the ROVs. Inferable from the considerable uncertainties and totally indefinite dynamics of the AUV, an effective NN law was used for the following control of AUVs [28]. Despite the fact that the previously mentioned critical control techniques have been accounted for control execution and scientists had the option to introduce great quality execution. However, essentially, these control methods are short of robustness in opposition to the uncertainties.

Hence, encouraged by this thought, the inspiration of this exertion is to propose a high-accuracy ensured NN-based hybrid backstepping scheme that consolidates the compensation of the model-reliant and the model-independent control strategies. For the direction following control issue of the AUVs, the essential attempt that straightforwardly connects the distance between the various version-reliant and the without model control strategies. Generally, it is considered that the model-independent controllers want no longer any earlier records regarding the dynamics of the scheme, although this isn't always the real case. In actual programs, we are continuously handy with restricted records about framework factors; therefore, it is continuously purposeful that it is essential to use these facts for the controller proposed system. This is the motivator for our evaluation. In this, the authors have included the model-based control method with the model-free technique for the layout of the neural network (NN)–based efficient backstepping control scheme for the AUVs. In this, the unknown dynamics is efficaciously estimated with the assistance of an NN. Furthermore, a compensator is also within the controller part for the reason that unidentified consequences such as the NN reconstruction error outdoors disturbances can also nonetheless make the scheme unstable. For the scrupulous evidence of the steadiness of the close-looped device, the Lyapunov theorem as well as Barbalat lemma is employed within the stability principle. As a result, errors in all the orders congregate to 0 asymptotically. Relative numerical effects are created for the workability, confirmation, and consistency of the proposed approach.

2.2 DYNAMICAL MODEL DESCRIPTION

In many works, the powerful model of the AUVs has been planned [31–33]. Because of the consideration of unmodeled elements and hydrodynamic components in the dynamical, the unique model turns out to be a lot of complex and profoundly non-linear (Figure 2.1). Accordingly, most importantly the summed-up powerful form of AUV is determined and later, for the practicability of the presented control method, a 4-DOF diminished unique model is inferred. To accomplish the kinematic form of the AUV frameworks, we typically consider inertial/earth-fixed casing and the body-fixed edge. In the body-fixed casing, $v = [\bar{u}, \ \bar{v}, \ \bar{w}, \ \bar{p}, \ \bar{q}, \ \bar{r}]^T$ indicate the velocities such that $\bar{u}, \ \bar{v}, \ \bar{w} \in R$ are the surge, sway, and heave linear

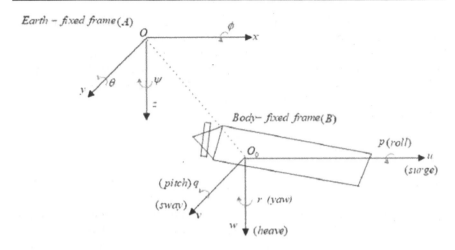

Figure 2.1 AUV coordinate system.

velocities and $\bar{p}$, $\bar{q}$, $\bar{r} \in R$ depict angular velocities. While in the earth-fixed casing, $\mu = [x, \; y, \; z, \; \emptyset, \; \theta, \; \Psi]^T$ is the positional and orientation vector to such an extent that $x, \; y, \; z \in R$ is the position of the vehicle and $\emptyset, \; \theta, \; \Psi \in R$ stand for the roll, pitch, and yaw. Subsequently, the kinematic conditions of movement is portrayed as follows:

$$\dot{\mu} = J_m(\mu)v \tag{2.1}$$

$$\ddot{\mu} = J_m(\mu)\dot{v} + \dot{J}_m(\mu)v \tag{2.2}$$

where the Jacobian transformation matrix $J_m(\mu) \in R^{6\times6}$ between the two frames is detailed as

$$J_m(\mu) = \begin{bmatrix} J_{m1}(\mu) & 0_{3\times3} \\ 0_{3\times3} & J_{m2}(\mu) \end{bmatrix}$$

$$J_{m1}(\mu) = \begin{bmatrix} cos\Psi cos\theta & sin\,\theta sin\emptyset cos\Psi - cos\emptyset sin\Psi & sin\,\emptyset sin\Psi + sin\theta cos\emptyset cos\Psi \\ sin\Psi cos\theta & sin\,\theta sin\emptyset sin\Psi + cos\emptyset cos\Psi & sin\,\theta cos\emptyset sin\Psi - sin\emptyset cos\Psi \\ -sin\,\theta & sin\emptyset cos\theta & cos\emptyset cos\theta \end{bmatrix}$$

$$J_{m2}(\mu) = \begin{bmatrix} 1 & sin\emptyset tan\theta & cos\emptyset tan\theta \\ 1 & cos\emptyset & -sin\emptyset \\ 0 & sin\emptyset sec\theta & cos\emptyset sec\theta \end{bmatrix}$$

Next, we suppose that the consequent hypothesis is used on the autonomous AUV structure as:

Assumption I: Jacobian matrix $J_m(\mu)$ inverse does not exist at $\theta = \pm\frac{\pi}{2}$, therefore, let $-\frac{\pi}{2} < \theta < \frac{\pi}{2}$.

Next, the dynamic model of AUV is introduced as

$$M_m \dot{v} + C_m(v)v + D_m(v)v + g_m(\mu) + T_{dm} = \tau_m \tag{2.3}$$

The framework $M_m \in R^{6 \times 6}$ is the inertia matrix of the AUV including the rigid body and the additional mass. Whereas $C_m(v) \in R^{6 \times 6}$ is the Coriolis and the centripetal matrix of the ith AUV including the rigid and the added mass terms. $D_m(v) \in R^{6 \times 6}$ addressing the hydrodynamic damping and the lift framework. $g_m(\mu) \in R^{6 \times 1}$ addressing the gravitational powers and the moments. $T_{dm} \in R^{6 \times 1}$ addressing the outer disturbances vector. The vector $\tau_m = [X, Y, Z, K, M, N]^T \in R^{6 \times 1}$ is a vector of summed-up powers applied on the AUV provided by the thrusters.

Next, from equation (2.1), the earth-fixed portrayal is acquired as

$$v = J_m^{-1}(\mu)\dot{\mu} \tag{2.4}$$

$$\dot{v} = J_m^{-1}(\mu)(\ddot{\mu} - \dot{J}_m(\mu)J_m^{-1}(\mu)\dot{\mu}) \tag{2.5}$$

After applying the kinematic transformation, we get

$$\overline{M}(\mu)\ddot{\mu} + \overline{C}(v, \mu)\dot{\mu} + \overline{D}(v, \mu)\dot{\mu} + \overline{g}(\mu) + \overline{T}_d = \overline{\tau}_d \tag{2.6}$$

where $\overline{M}_m = J_m^{-T}(\mu)M_m J_m^{-1}(\mu)$; $\overline{C}(v, \mu) = J_m^{-T}(\mu)[C_m(v) - M_m J_m^{-1}(\mu)\dot{J}_m(\mu)]$ $J_m^{-1}(\mu)$; $\overline{D}(v, \mu) = J_m^{-T}(\mu)D_m J_m^{-1}(\mu)$; $\overline{g}(\mu) - J_m^{-T}g_m$; $\overline{T}_d = J_m^{-T}T_{dm}$

Property I: The matrix $\overline{M}(\mu)$ is symmetric as well as positive definite.
Property 2: The matrix skew–symmetric relationship is satisfied between inertia and Coriolis matrix.
Property 3: External disturbances are bounded.

2.3 ERROR DYNAMICS CONTROLLER DESIGN

The control idea of the work is to suggest an NN-based hybrid backstepping powerful control scheme for direction following management of AUV by making use of model-based (MB), model-free (MF), and the conventional backstepping approaches. with solid powerfulness contrary to uncertainties. The introduced control strategy ought to have the option to make certain that the tracking inaccuracies unite to 0 and the entire proscribed framework is asymptotically steady. Consider $\mu_d = [x_d, \ y_d, \ z_d, \ \Psi_d]^T$; the desired vector and define the tracking error and filtered errors as

$$\bar{\mu} = \mu_d - \mu \tag{2.7}$$

$$\mu = \dot{\bar{\mu}} + K_\mu \bar{\mu} \tag{2.8}$$

where K_μ is a positive constant matrix. After that, the desired velocity vector is described by

$$v_d = J_m^{-T}(\mu_d)\dot{\mu}_d \tag{2.9}$$

such that after taking the help of kinematic equation (2.1), the velocity tracking error can be attained

$$\bar{v} = v_d - v \tag{2.10}$$

After differentiating (2.8) and using (2.7), the dynamic equation (2.6) in the form µ can be established as

$$\overline{M}_m \dot{\rho} + \overline{C}_m \rho = G(n) - \bar{\tau}_m + \overline{T}_d \tag{2.11}$$

where $\quad G(n) = \overline{M}_m(\ddot{\mu}_d + K_\mu \dot{\bar{\mu}}) + \overline{D}_m \dot{\mu} + \bar{g}_m \quad$ with the input vector $n = [\bar{\mu}, \ \dot{\bar{\mu}}, \ \mu_d, \ \ddot{\mu}_d]$.

As the AUVs are exposed to hydrodynamic burdens, hence, it is a tough undertaking to estimate and compute the hydrodynamic factors that are fundamental for the vehicle's working situations. As such, no preceding statistics approximately the framework's dynamics is exactly recognized. In this manner, whatever data about framework elements exists ought to be utilized for the design of the controller. Accordingly, we split non-linear mechanical capacity into two sections; $\hat{G}(n)$ the identified dynamic part and $\tilde{G}(n)$ the obscure powerful part like:

$$G(n) = \hat{G}(n) + \tilde{G}(n) \tag{2.12}$$

where $\quad \hat{G}(n) = \widehat{\overline{M}}_m(\ddot{\mu}_d + K_\mu \dot{\bar{\mu}}) + \widehat{\overline{C}}_m(\dot{\mu}_d + K_\mu \bar{\mu}) + \hat{D}_m \dot{\mu} + \widehat{\bar{g}}_m \quad$ and $\quad \tilde{G}(n) = \widetilde{\overline{M}}_m(\ddot{\mu}_d + K_\mu \dot{\bar{\mu}}) + \widetilde{\overline{C}}_m(\dot{\mu}_d + K_\mu \bar{\mu}) + \widetilde{D}_m \dot{\mu} + \widetilde{\bar{g}}_m$

To this cease, we use characteristic RBFNN [34] for the unknown dynamic component's approximation.

$$\tilde{G}(n) = \overline{H}^T \kappa(n) + \epsilon(n) \tag{2.13}$$

where $\overline{H} \in R^{s \times \overline{y}}$ is an ideal weight matrix, n represents the input vector. $\epsilon(\cdot)$: $R^{5\overline{p}} \to R^S$ representing the approximation error fulfilling $\|\epsilon(n)\| < \epsilon_S$ for $\epsilon_S > 0$. $\kappa(n)$ is the Gaussian feature mathematically set up as: $\kappa(n) = \exp\left(\dfrac{-\|n - e_j\|^2}{2g_j^2}\right)$. Here, e_j and g_j represent the centres and the widths of RBFNN, respectively. More specifically, the factors of $\omega_j(n)$ are fulfilling $0 < \kappa_j(n) < 1$.

After substituting $\tilde{G}(n)$ from equation (2.13) in equation (2.11), the changed equation is as follows:

$$\overline{M}_m \dot{\rho} + \overline{C}_m \rho = \hat{G}(n) + \overline{H}^T \kappa(n) + \epsilon(n) + \overline{T}_d - \overline{\tau}_m \tag{2.14}$$

Next, we recommend the subsequent hybrid controller for AUV as

$$\overline{\tau}_m = \hat{G}(n) + B_d \rho + \widehat{\overline{H}}^T \kappa(n) + \overline{\mu} + \varDelta \tag{2.15}$$

where $\widehat{\overline{H}}$ is added because of the approximation of NN weights. $B_d = B_d^T$ and $\varDelta$ are the gain matrix and the adaptive term, respectively. Here, to deactivate the impacts of NNs estimation error and the affects of the unsettling impacts, Δ is delivered to the regulator part that's deliberate as follows:

Later than employing assets 3 and call upon $\|\epsilon(n)\| < \epsilon_S$, the subsequent inequality is set up:

$$\|\overline{T}_d + \epsilon(n)\| \leq t_1 + \epsilon_S \tag{2.16}$$

We have as follows

$$\overline{A} = t_1 + \epsilon_S \tag{2.17}$$

which can be given in equation (2.18) in representative form as

$$\overline{A} = [1 \ \ 1][t_1 \ \ \epsilon_S]^T = \overline{Z}^T \overline{X} \tag{2.18}$$

In view of equations (2.16), (2.17), and (2.18) and using $\dot{\overline{A}} = -\Upsilon\overline{A}$ with $\overline{A}(0) > 0$; a consistent design, Δ is delivered in the form

$$\Delta = \frac{\widehat{\widehat{A}}^2 \rho}{\widehat{\widehat{A}} \, \|\rho\| + \overline{\Delta}} \tag{2.19}$$

For the weighted network and the parameter vector, the accompanying adaptive control laws are as follows:

$$\dot{\widehat{\widehat{H}}} = \Gamma_{\widehat{H}} \kappa(n) \rho^T \tag{2.20}$$

$$\dot{\widehat{\widehat{X}}} = \Gamma_{\widehat{X}} \, \overline{Z} \|\rho\| \tag{2.21}$$

where $\Gamma_{\widehat{H}} = \Gamma_{\widehat{H}}^T \in R^{f_1 \times f_1}$ and $\Gamma_{\widehat{X}} = \Gamma_{\widehat{X}}^T \in R^{f_2 \times f_2}$ should be chosen the matrices of positive definiteness.

Now taking the equations (2.19) and (2.15) together and using in equation (2.14), the error dynamics is converted in the form:

$$\overline{M}_m \dot{\rho} + \overline{C}_m \rho = -B_d \rho - \overline{\mu} + \widetilde{\widehat{H}}^T \kappa(n) - \frac{\widehat{\widehat{A}}^2 \rho}{\widehat{\widehat{A}} \, \|\rho\| + \overline{\Delta}} + \epsilon(n) + \overline{T}_d \tag{2.22}$$

2.4 STABILITY ANALYSIS

To show the steadiness of the overall structure and the robustness of the proposed control scheme, we outline the Lyapunov characteristic as a feature of the subsequent variables:

$$\tilde{L}_p = \frac{1}{2} \rho^T \overline{M}_m \rho + \frac{1}{2} tr \left(\widetilde{\widehat{H}}^T \Gamma_{\widehat{H}}^{-1} \widetilde{\widehat{H}} \right) + \frac{1}{2} tr \left(\widetilde{\widehat{X}}^T \Gamma_{\widehat{X}}^{-1} \widetilde{\widehat{X}} \right) + \frac{\overline{\Delta}}{\gamma} \tag{2.23}$$

Differentiation of equation (2.23) gives

$$\dot{\tilde{L}}_p = \frac{1}{2} \rho^T \dot{\overline{M}}_m \rho + \rho^T \overline{M}_m \dot{\rho} + tr \left(\widetilde{\widehat{H}}^T \Gamma_{\widehat{H}}^{-1} \dot{\widetilde{\widehat{H}}} \right) + \frac{1}{2} tr \left(\widetilde{\widehat{X}}^T \Gamma_{\widehat{X}}^{-1} \dot{\widetilde{\widehat{X}}} \right) + \frac{\dot{\overline{\Delta}}}{\gamma} \tag{2.24}$$

$$\dot{\tilde{L}}_p = \frac{1}{2} \rho^T \dot{\overline{M}}_m \rho + \rho^T [-B_d \rho - \overline{\mu} + \widetilde{\widehat{H}}^T \kappa(n) + \epsilon(n) + \overline{T}_d - \overline{C}_m \mu] - \frac{\widehat{\widehat{A}}^2 \rho \rho^T}{\widehat{\widehat{A}} \, \|\rho\| + \overline{\Delta}}$$
$$+ tr \left(\widetilde{\widehat{H}}^T \Gamma_{\widehat{H}}^{-1} \dot{\widetilde{\widehat{H}}} \right) + tr \left(\widetilde{\widehat{X}}^T \Gamma_{\widehat{X}}^{-1} \dot{\widetilde{\widehat{X}}} \right) - \overline{\Delta} \tag{2.25}$$

Using the relations $\dot{\widetilde{H}} = -\dot{\widehat{H}}$ and $\dot{\widetilde{X}} = -\dot{\widehat{X}}$, we get

$$\dot{L}_p = \frac{1}{2}\rho^T(\dot{\overline{M}}_m - 2\overline{C}_m)\rho - \rho^T B_d\rho - \rho^T\overline{\mu} + \rho^T\widetilde{H}^T\kappa(n) + \rho^T(\overline{T}_d + \epsilon(n))$$
$$- \frac{\widehat{A}^2\rho\rho^T}{\widehat{A}\|\rho\| + \overline{A}} - tr\left(\widetilde{H}^T\Gamma_{\overline{H}}^{-1}\dot{\widetilde{H}}\right) - tr\left(\widetilde{X}^T\Gamma_{\overline{X}}^{-1}\dot{\widetilde{X}}\right) - \overline{A} \qquad (2.26)$$

Now consider,

$$\rho^T(\overline{T}_d + \epsilon(n)) \leq \|\mu\rho\|\|(\overline{T}_d + \epsilon(n))\| \leq \overline{A}\|\rho\| = Z^T(\widehat{X} + \widetilde{X})\|\rho\| \qquad (2.27)$$

According to Property 2 and inequality (2.27), one can easily write the equation (2.26) in (2.27) as:

$$\dot{L}_p \leq -\rho^T B_d\rho - \rho^T\overline{\mu} + Z^T(\widehat{X} + \widetilde{X})\|\rho\| + \rho^T\widetilde{H}^T\kappa(n) - tr\left(\widetilde{H}^T\Gamma_{\overline{H}}^{-1}\dot{\widetilde{H}}\right)$$
$$- tr\left(\widetilde{X}^T\Gamma_{\overline{X}}^{-1}\dot{\widetilde{X}}\right) - \overline{A} - \frac{\widehat{A}^2\rho\rho^T}{\widehat{A}\|\rho\| + \overline{A}} \qquad (2.28)$$

According to the weight and parameter law, we have as follows:

$$\dot{L}_p \leq -\rho^T B_d\rho - \rho^T\overline{\mu} + Z^T(\widehat{X} + \widetilde{X})\|\rho\| + \rho^T\widetilde{H}^T\kappa(n) - tr(\widetilde{H}^T\kappa(n)\rho^T)$$
$$- tr(\widetilde{X}^T\overline{Z}\|\rho\|) - \overline{A} - \frac{\widehat{A}^2\rho\rho^T}{\widehat{A}\|\rho\| + \overline{A}} \qquad (2.29)$$

Simplification of equation (2.29) results in equations as follows:

$$\dot{L}_p \leq -\rho^T B_d\rho - \rho^T\overline{\mu} + \frac{(\overline{Z}^T\widehat{X})\|\rho\|\overline{A}}{(\overline{Z}^T\widehat{X})\|\rho\| + \overline{A}} - \overline{A} \qquad (2.30)$$

$$\dot{L}_p \leq -\rho^T B_d\rho - \rho^T\overline{\mu} - \frac{\overline{A}^2\|\rho\|^2}{(\overline{Z}^T\widehat{X})\|\rho\| + \overline{A}} \leq -\rho^T B_d\rho \qquad (2.31)$$

$$\ddot{L}_p \leq -2\rho^T B_d\dot{\rho} \qquad (2.32)$$

By using the application of Barbalat's lemma [35] point out that $\dot{L}_p$ converges to zero as $t \to \infty$ and the overall system is asymptotically stable.

2.5 SIMULATION STUDY

Some simulation examples have been developed in MATLAB to prove the efficiency and workability of the hybrid backstepping powerful control scheme. Here, we will apply the proposed hybrid control scheme over a simplified model of AUV with 4-DOF AUV, i.e., control in x, y, z, and $\overline{\Psi}$ directions.

Case 1: In this, we evoke the preferred circular route as:

$$x_d = 10\sin(0.01t); \quad y_d = 10\cos(0.01t); \quad z_d = 10; \quad \Psi_d = \frac{\pi}{3} \quad (2.33)$$

In the simulation study, we have utilized two diverse control techniques including the existing control scheme point by point in [36], and the proposed control stratagem addressed. The advancement of the position and velocity tracking inaccuracies as well as forces and torques are displayed in Figures 2.2–2.7 for case 1 with the existing technique and the proposed law. From these figures, it is obvious that the dynamic compensation cannot be properly handled appreciably by [36].

Tracking of AUV in circular path under the proposed control law

Case2: $x_d = 10\sin(0.01t); \quad y_d = 10\cos(0.01t); \quad z_d = 10t; \quad \Psi_d = \frac{\pi}{3}$ \quad (2.34)

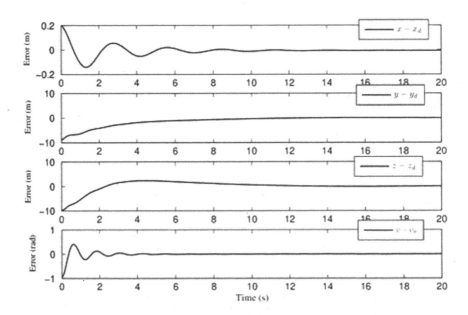

Figure 2.2 Position inaccuracies by Sahu and Bidyadhar [36] for a circular path.

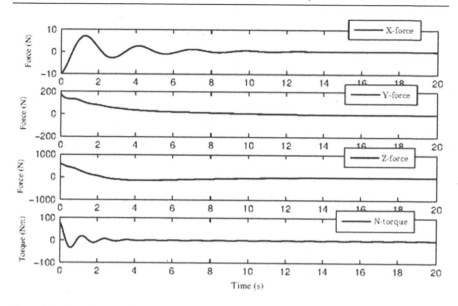

Figure 2.3 Circular path forces and torques by Sahu and Bidyadhar [36].

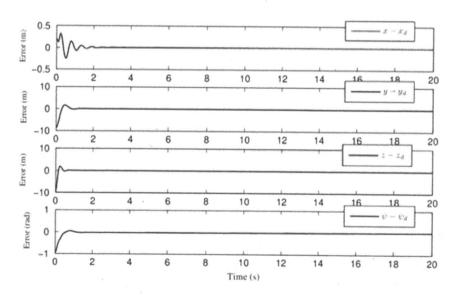

Figure 2.4 Circular path position inaccuracies under the proposed control technique.

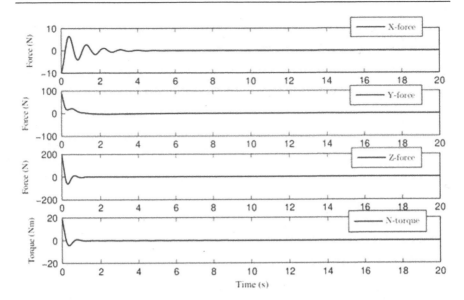

Figure 2.5 Circular path forces and torques by the proposed technique.

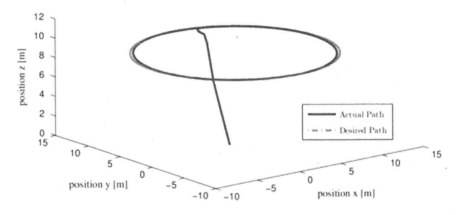

Figure 2.6 Tracking of AUV in the round path by Sahu and Bidyadhar [36].

The development of the position and velocity tracking errors as well as forces and torques are shown in Figures 2.8–2.15 for the case 2 with the existing technique and the proposed law.

From the figures, it is obvious that due to the fact the to be had dynamic information has been implemented successfully by means of way of the advanced manipulate method. Finally, from the above discussion, it's far seen that the proposed method completely takes assist of the model-reliant

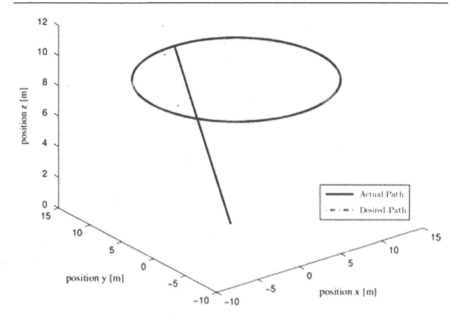

Figure 2.7 Tracking of AUV in the round path under the proposed method.

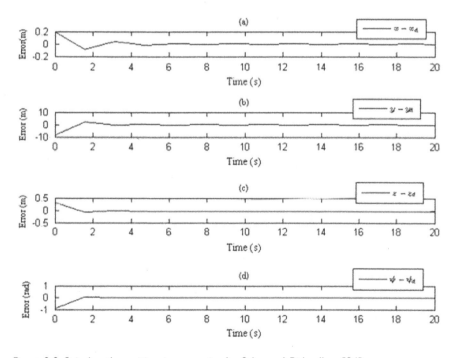

Figure 2.8 Spiral path position inaccuracies by Sahu and Bidyadhar [36].

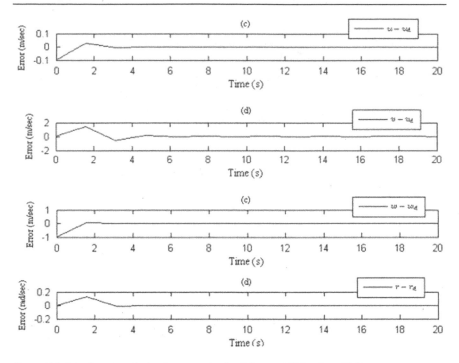

Figure 2.9 Spiral path velocity inaccuracies by Sahu and Bidyadhar [36].

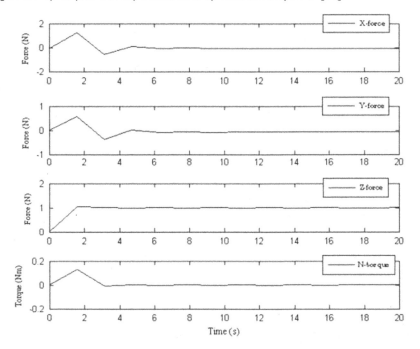

Figure 2.10 Forces and torques inside the spiral route by Sahu and Bidyadhar [36].

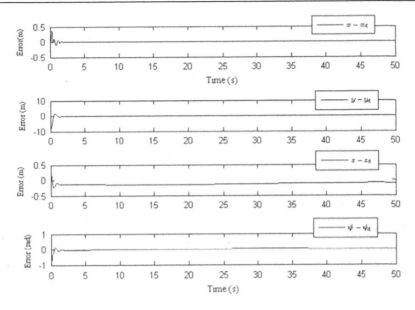

Figure 2.11 Spiral path position inaccuracy under the proposed method.

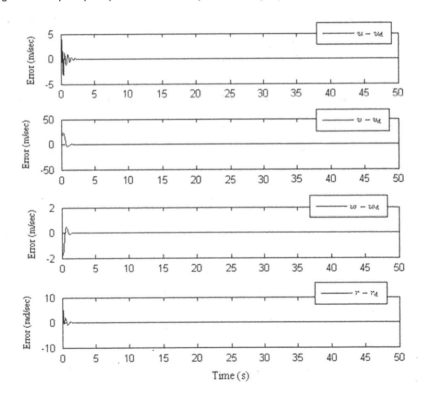

Figure 2.12 Spiral path velocity inaccuracy under the proposed method.

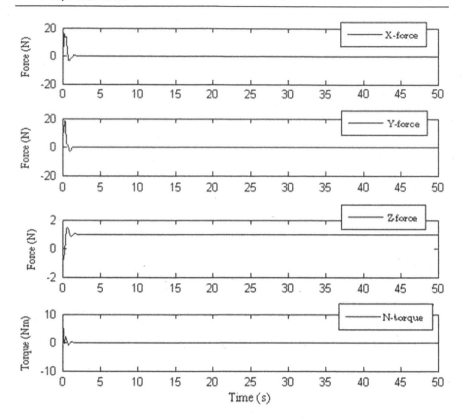

Figure 2.13 Forces and torques inside the spiral route by the proposed control law.

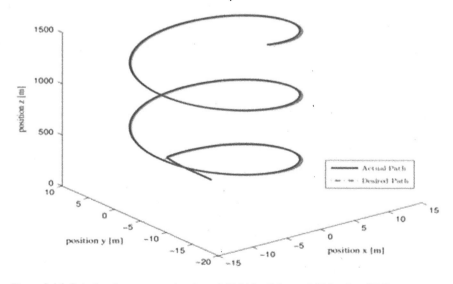

Figure 2.14 Spiral trajectory monitoring of AUV by Sahu and Bidyadhar [36].

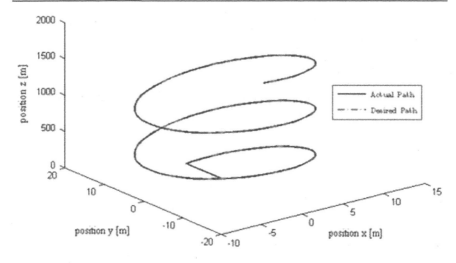

Figure 2.15 Spiral trajectory monitoring of AUV under the proposed method.

and the model-independent manipulation techniques as it should be and nonetheless a lot less induced with the useful resource of the uncertainties.

REFERENCES

[1] Sahoo, A., Dwivedy, S. K., & Robi, P. S. (2019). Advancements in the field of autonomous underwater vehicle. *Ocean Engineering*, *181*, 145–160.
[2] Puall, L., Saeedi, S., Seto, M., & Li, H. (2014). AUV navigation and localization: A review. *IEEE Journal of Ocean Engineering*, *39*, 131–149.
[3] Qu, Y., Xiao, B., Fu, Z., & Yuan, D. (2018). Trajectory exponential tracking control of unmanned surface ships with external disturbance and system uncertainties. *ISA Transactions*, *78*, 47–55.
[4] Yu-lei, L., Ming-Jun, Z., Lei, W., & Ye, L. (2018). Trajectory tracking control for underactuated unmanned surface vehicles with dynamic uncertainties. *Journal of Central South University*, *23*, 370–378.
[5] Rezazadegana, F., Shojaei, K., Sheikholeslam, F., & Chatraei, A. (2015). A novel approach to 6-DOF adaptive trajectory tracking control of an AUV in the presence of parameter uncertainties. *Ocean Engineering*, *107*, 246–258.
[6] Zheng, Z., Ruan, L., & Zhu, M. (2019). Output-constrained tracking control of an underactuated autonomous underwater vehicle with uncertainties. *Ocean Engineering*, *175*, 241–250.
[7] Lamraoui, H. C., & Qidan, Z. (2019). Path following control of fully-actuated autonomous underwater vehicle in presence of fast-varying disturbances. *Applied Ocean Research*, *86*, 40–49.
[8] Liu, F., Shen, Y., He, B., Wang, D., Wan, J., Sha, Q., & Qin, P. (2019). Drift angle compensation-based adaptive line-of-sight path following for autonomous underwater vehicle. *Applied Ocean Research*, *93*.

[9] Peng, Z., Wang, J., & Han, Q. L. (2018). Path-following control of autonomous underwater vehicles subject to velocity and input constraints via neurodynamic optimization. *IEEE Transactions on Industrial Electronic*, 66, 8724–8732.

[10] Yan, Z., Yu, H., Zang, W., Li, B., & Zhou, J. (2015). Globally finite-time stable tracking control of underactuated UUVs. *Ocean Engineering*, 107, 132–146.

[11] Maurya, P., Desa, E., Pascoal, A., Barros, E., Navelkar, G., Madhan, R., et al. (2019). Control of the Maya AUV in the vertical and horizontal planes: theory and practical results. *7th IFAC Conference on Maneuvring and Control of Marine Craft*, 1–7.

[12] Hassanein, O., Anavatti, S. G., Shim, H., & Ray, T. (2016). Model-based adaptive control system for autonomous underwater vehicles. *Ocean Engineering*, 127, 58–69.

[13] Smallwood, D. A., & Whitcomb, L. L. (2004). Model-based dynamic positioning of underwater robotic vehicles: Theory and experiment. *IEEE Journal of Oceanic Engineering*, 29, 169–186.

[14] Guerrero, J., Torres, J., Creuze, V., & Chemori, A. (2019). Trajectory tracking for autonomous underwater vehicle: An adaptive approach. *Ocean Engineering*, 172, 511–522.

[15] Li, J., Du, J., & Chang, W. J. (2019). Robust time-varying formation control for underactuated autonomous underwater vehicles with disturbances under input saturation. *Ocean Engineering*, 179, 180–188.

[16] Xu, J., Wang, M., & Qiao, L. (2015). Dynamical sliding mode control for the trajectory tracking of underactuated unmanned underwater vehicles. *Ocean Engineering*, 105, 54–63.

[17] Yan, Y., & Yu, S. (2018). Sliding mode tracking control of autonomous underwater vehicles with the effect of quantization. *Ocean Engineering*, 151, 322–328.

[18] Elmokadem, T., Zribi, M., & Youcef-Toumib, K. (2017). Terminal sliding mode control for the trajectory tracking of underactuated autonomous underwater vehicles. *Ocean Engineering*, 129, 613–625.

[19] Qiao, L., & Zhang, W. (2019). Double-loop integral terminal sliding mode tracking control for UUVs with adaptive dynamic compensation of uncertainties and disturbances. *IEEE Journal of Oceanic Engineering*, 44, 29–53.

[20] Qiao, L., & Zhang, W. Trajectory tracking control of AUVs via adaptive fast nonsingular integral terminal sliding mode control. *IEEE Transactions on Industrial Informatics*. 10.1109/TII.2019.2949007.

[21] Yan, Z., Wang, M., & Xu, J. (2019). Robust adaptive sliding mode control of underactuated autonomous underwater vehicles with uncertain dynamics. *Ocean Engineering*, 173, 802–809.

[22] Sun, B., Zhu, D., & Li, W. (2012). An integrated backstepping and sliding mode tracking control algorithm for unmanned underwater vehicles. *UKACC International Conference on Control*, 644–649.

[23] Li, J. H., & Lee, P. M. (2005). Design of an adaptive nonlinear controller for depth control of an autonomous underwater vehicle. *Ocean Engineering*, 32, 2165–2181.

[24] Elhaki, O., & Shojaei, K. (2018). Neural network-based target tracking control of underactuated autonomous underwater vehicles with a prescribed performance. *Ocean Engineering, 167,* 239–256.

[25] Venugopal, K. P., Pandya, A. S., & Sudhakar, R. (1994). A recurrent neural network controller and learning algorithm for the on-line learning control of autonomous underwater vehicles. *Neural Networks, 7,* 833–846.

[26] Li, J. H., Lee, P. M., & Lee, S. J. (2002) *Motion Control of an AUV Using a Neural Network Adaptive Controller.* USA: IEEE.

[27] Bagheri, A., & Moghaddam, J. J. (2009). Simulation and tracking control based on neural-network strategy and sliding-mode control for underwater remotely operated vehicle. *Neurocomputing, 72,* 1934–1950.

[28] Pan, C. Z., Lai, X. Z., Yang, S. X., & Wu, M. (2013). An efficient neural network approach to tracking control of an autonomous surface vehicle with unknown dynamics. *Expert Systems with Applications, 40,* 1629–1635.

[29] Xu, B., Pandian, S. R., Sakaggami, N., & Petry, F. (2012). Neuro-fuzzy control of underwater vehicle-manipulator systems. *Journal of the Franklin Institute, 349,* 1125–1138.

[30] Chen, Y., Wang, K., & Chen, W. (2019). Adaptive fuzzy depth control with trajectory feedforward compensator for autonomous underwater vehicles. *Advances in Mechanical Engineering, 11*(3), 1–12.

[31] Fossen, T. I. (1994). *Guidance and Control of Ocean Vehicles.* John Wiley & Sons Inc.

[32] Do, K. D., & Pan, J. (2009). Control of ships and underwater vehicles: Design for underactuated and nonlinear marine systems. *Springer Science and Business Media, 11,* 210–224.

[33] Antonelli, G., Fossen, T. L., & Yogereger, D. R. (2008). *Underwater Robotics in Springer Handbook of Robotics.* USA: Springer, pp. 987–1008.

[34] Park, J., & Sandberg, J. W. (1991). Universal approximation using radial basis function networks. *Neural Computations, 3,* 246–257.

[35] Slotine, J. J. E., & Li, W. (1991). *Applied Nonlinear Control.* Singapore: Prentice-Hall.

[36] Sahu, B. K., & Bidyadhar, S. (2014). Adaptive tracking control of autonomous underwater vehicle. *International Journal of Automation and Computing, 11*(3), 299–307.

Chapter 3

Integrated Storage Allocation and Privacy Preserve Deduplication Process over Cloud Using Dekey Method

*F. Antony Xavier Bronson, V. Sai Shanmuga Raja,
Saravanan Elumalai, G. Senthil Velan, and P. Solainayagi*

CONTENTS

DOI: 10.1201/9781003355960-3

3.1 INTRODUCTION

3.1.1 Cloud Computing

Cloud computing can be defined as a service-based computer architecture wherein cloud customers are provided with services based on their needs [1]. Internet is a backbone to provide delivering high bandwidth on need. The services are dynamically reliant on virtual machine resources [2]. The cloud computing paradigm has a number of qualities, which are outlined as services that are extremely scalable, computing resources with a high degree of flexibility, reliability, availability, easily transportable, pay-as-you-go is a way of operation that allows you to pay as you go, resources that have been optimized, maximum efficiency, and interact with a variety of services [3]. As a result, it's vital to use an efficient technique for managing memory usage in cloud networks. Current strategies for resource allocation on cloud platforms use probability distribution techniques to make effective use of available memory, yet accessible memory is not used in cloud platforms. The host flexibility standard has now been adopted for memory management and memory balancing within the host [4,5].

3.1.2 Utility Computing

Utility computing is the cloud computing cost of a service system that relies on computation resources and infrastructure design that is managed by the provider. The most efficient method at the lowest cost is often used in this system. It follows the cloud computing metering service with system resources such as processing, service-based storage, and so on. It has the metering-based bandwidth management, self-sufficiency, and exceptional scalability characteristics.

3.1.3 Web Service

Service is the name of the function that has the abstraction property. With connection endpoints, web services connect distinct services. The service performs a variety of functions that are fully defined, self-contained, and self-contained in nature. Web services are self-descriptive and stateless in nature, allowing them to perform autonomous tasks.

3.1.4 Web Service Description Language

Web service description language (WSDL) is an XML-based description language for describing services on a remote system. Customer, service provider, and registry are the three main components. Always, the customer generates the request by consulting the WSDL description. The service provider is responsible for offering services to customers depending on their needs. The registry acts as a go-between storing XML descriptions of the services.

3.1.5 Architecture of Web Services

The service provider, service user, and service registry are the three main components of a web service. At their end, the service provider maintains the service implementation. The service is provided using an open standard and takes the form of message processing in response to user requests. The registry stores the description of the service implementation. It keeps track of the different pages depending on their usage and qualities. Through the register, the user learns about the available services from the provider. Once the service has been recognized, messages are transmitted. Simple object access protocol (SOAP) is used to address this. Figure 3.1 depicts the web service's conceptual diagram.

3.1.6 Service-Oriented Architecture

A service-oriented architecture (SOA) is defined as a set of self-contained services that interact with one another. It has characteristics such as

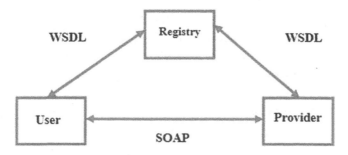

Figure 3.1 Web service interaction.

Figure 3.2 Consumer–provider interaction.

autonomy, interoperability, scalability, and reusability. The message exchange pattern is used to trade the services. The SOA has several layers, including service analysis, service design, and service implementation. Figure 3.2 depicts the SOA service interaction.

3.1.7 Service Excellence

Quality of service (QoS) characteristics are used to evaluate network traffic, service latency, and efficiency, among other things. This also examines user expectations based on a variety of experiences with personal and business networks. The focus is mostly on client retention and technological adoption.

3.1.8 Service Level Agreement

A service level agreement (SLA) is indeed an agreement between a service user and a service provider which allows them to combine several services. Various organizations maintain their own service descriptions, which are insufficient to meet the needs of clients.

3.1.9 Factors Affecting Performance Design

In order to provide clients with continuous services, scalability refers to the ability to improve the performance of network elements, communications networks, process-related conversations, software, and hardware. Elasticity is a technique for increasing the quantity of resources attached to a server. Computing resources such as servers, networks, and clients, as well as their management, were treated like instances in dynamic provisioning. *Availability*: In cloud and distributed systems, availability has been utilized to ensure that certain resources are always available. A single system or a cluster of systems and subsystems can be used.

3.1.10 Fault-Tolerant System

A fault-tolerant system is one that continues to work for any process without interruption, even if one or more of its components fails. There are

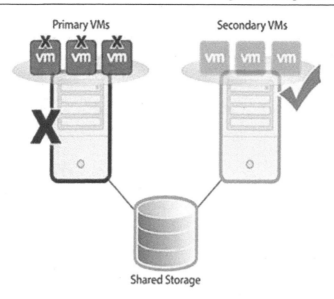

Figure 3.3 Fault-tolerant systems.

many other qualities and attributes, such as robustness, security, and so on. The conceptual diagram of a fault tolerance system is shown in Figure 3.3.

3.1.11 Back Up Your Data

A data backup is a method of preserving a duplicate copy of the original data. Full backups, differential backups, and incremental backups are all examples of backup methods. This is mostly used for real-time data storage systems [6,7]. *Cloud Storage Backup*: Cloud storage makes use of backup as a service (BaaS). This service is built on a subscription model and includes characteristics such as cheap cost, scalability, capacity expansion, data redundancy elimination, and hardware backup. To protect data integrity, the data are encrypted before being stored in the cloud storage. The three types of cloud backup are hybrid cloud storage, private cloud storage, and public cloud storage. The cloud backup mechanism is depicted in Figure 3.4.

3.1.12 Data Streaming

Data streaming is the continuous transmission of data from numerous real-time sources. This necessitates a plethora of performance characteristics, including bandwidth and storage media. A data stream's performance is the ability to move data to the appropriate devices with little latency. The process of the data stream from multiple sources is depicted in Figure 3.5.

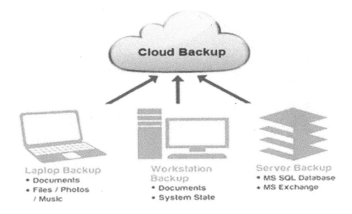

Figure 3.4 Cloud-based backup services.

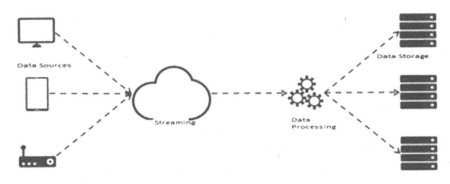

Figure 3.5 Data streaming process.

3.1.13 Data from Transactions

SAP and financial applications employing transaction processing systems (TPS) are two examples of applications that are used to process transactional data. In order to sustain the transaction retention rate, record management is critical. The data warehousing approach is used to assess the summarization of data over transaction data [8]. The data organization is depicted in Figure 3.6.

3.1.14 Extraction of Data

Using migration, import, and export capabilities, the data are processed and stored in a huge collection of clusters. Meta-data is also kept in order to make it easier to understand the data [9]. The data extraction procedure has been used on a computer as well as any other device that can access the data. Pattern matching and regular expression-based analysis are used to identify large amounts of data. The data extraction method is depicted in Figure 3.7.

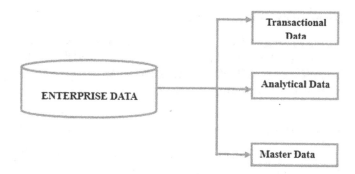

Figure 3.6 Organization of data.

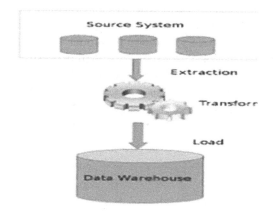

Figure 3.7 Data extraction process.

3.1.15 Cloud Protection

While transferring data from the location of a service provider to the location of users, it has been protected using a variety of safety measures to increase dependability. Three types of security in the computer platform: computing security, network security, and information security. The policies, technologies, data control, application protection, and hardware maintenance are all used to gauge security [10].

3.1.16 Problems with Cloud Management

Existing computing strategies govern and perform CPU-intensive tasks that take precedence over all others because the process leads to greater resource usage [11]. As a large range of providers share cloud services, there are chances of performance issues arising owing to a lack of authority over the resources. This could be controlled by assessing unique cloud services for a decent solution with small businesses to large businesses.

3.1.17 Computer Parallelism

The parallel processing software is based on the open MP and MPI programming models. The distributed computing model is depicted in Figure 3.8, while the parallel computing model is depicted in Figure 3.9. Figure 3.10 depicted the ubiquitous device interactions.

3.1.18 Distribution of Load

This is a technique for sharing the resources and work of multiple computers, as well as connectivity, hard disc drives, and other components that help with performance. The larger jobs are postponed until the overburdened processor completes its work. Smaller tasks are awaiting computation from a lightweight processor and an overload processor. Because no job is ready for execution, the idle processor always waits for it. The conceptual diagram for the load balancing model is shown in Figure 3.11. Handling of heavy workload processes, overload processors, and underload processors are the three types of load balancing algorithms. Load balancing uses a variety of algorithms to ensure that resources are available at all times during the computation process.

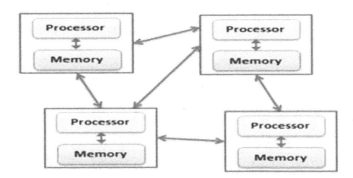

Figure 3.8 Distributed computing model.

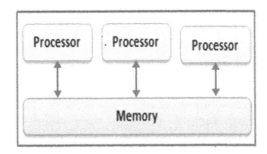

Figure 3.9 Parallel computing model.

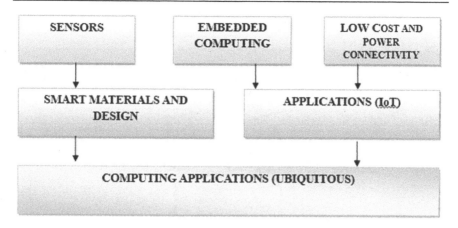

Figure 3.10 Interactions between ubiquitous.

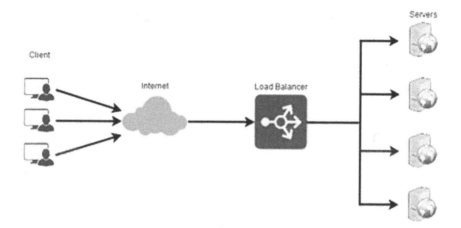

Figure 3.11 Load balancing conceptual diagram.

3.1.19 Scheduling of Tasks

The job scheduler is used to organize the jobs that are submitted for processing via batch processing from various customers. Distributed computing uses a scheduler to improve performance and has a variety of computing models based on their sector of interest. The major purposes of a scheduler are to schedule CPU-intensive tasks, grow and shrink tasks based on demand, and handle complex tasks using a pre-defined workflow. The procedure varies depending on the computing model. System maintenance is also done using dynamically configurable workflow management. Figure 3.12 depicts the cloud computing model's job scheduling.

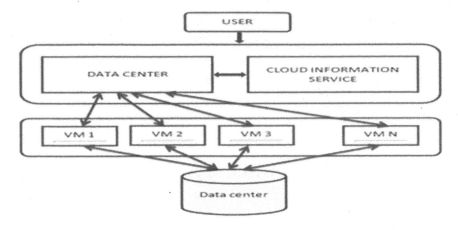

Figure 3.12 A scheduler for cloud computing.

3.1.20 Model of Cloud Service

Cloud computing is a computer model that provides services via the Internet that are ubiquitous and demand-oriented. Sharing networks, storage, data centres, and their services as provisioned through trustworthy interaction is how computing resources are configured. In the IT sector, there are three key service models namely Software as a Service (SaaS), Platform as a Service (PaaS), and Infrastructure as a Service (IaaS). The cloud service paradigm is depicted in Figure 3.13.

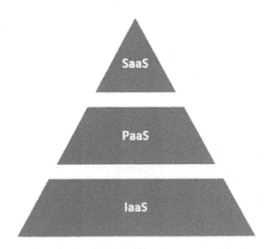

Figure 3.13 Models of cloud service.

3.1.21 Cloud Deployment Model

In the cloud, there seem to be four main models: public, private, community, and hybrid cloud. The term "public" is used to provide functions to cloud clients over a public network connection to the Internet. The cloud is owned by a third-party supplier with significant resources. The services are promoted to a broad variety of customers based on their requirements. Because the cloud manages data security, there is no comprehensive information about client data once it has been uploaded to cloud storage. The cloud provider is only responsible for the management of IT resources in terms of virtual machines [12]. Figure 3.14 depicts the cloud deployment models.

3.1.22 Deduplication of data

Data deduplication is a technology that stores data only once, ensuring that the same data is not duplicated across many cloud storage areas [13]. Data deduplication is a technique for reducing cloud storage space and maximizing bandwidth, the most complementary and growing challenge is secure deduplication. Although convergent key encryption is widely used, the Dekey and the new Dekey are widely used for safe data deduplication [14]. While migrating data across cloud service providers (CSPs), maintaining data privacy is a difficult process. Existing solutions do not protect against the disclosure of duplicate data. Pre- and post-process deduplication are the two types of deduplication processes [15]. The duplication information is removed because the processed data are kept in the storage, the pre deduplication procedure, which removes the duplication information, i.e., it blocks before the cloud storage, which provides optimal performance [16].

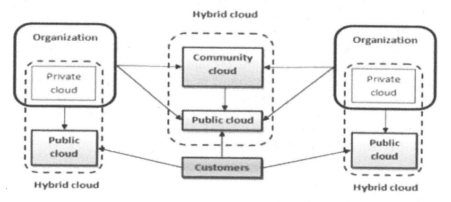

Figure 3.14 Model for cloud deployment.

3.1.23 Analytics on Big Data

Nowadays, data are collected from a variety of sources and stored in massive amounts. The practice of evaluating data from big data with related technologies is known as data analytics. Big data are a notion for storing large amounts of data in a way that makes the information more accurate and available to anybody who uses the internet. Hadoop uses the Map Reduce Programming model to store data named Hadoop Distributed File System. For large-scale real-time data sets, the Hadoop approach is ineffective. This difficulty is solved by utilizing cloud computing and numerous features that mimic the physical machine (PM). Cloud storage is the most important aspect as it stores large amounts of data without the need for clustering diverse computers.

3.1.24 Inspiration

Because of social media and the advancement of new technology, data are being created from a variety of sources. The information gathered is saved on the server. Because massive volumes of data arrive at a quick rate, storage server management is a demanding undertaking. There is a demand for server storage capacity, which drives up costs for both the provider and the consumer. The privacy areas protect data against unwanted disclosure by unknown people inside the region.

3.2 REDUCE DATA DEDUPLICATION IN CLOUD STORAGE WITH AN ENHANCED DEKEY APPROACH

This section explains the security algorithm that prevents customer data from being lost between the cloud provider and the cloud customer. In order to attain high dependability, many encryption methods are evaluated and compared to the proposed algorithms. The suggested dekeying method employs convergent keys that are dispersed across many cloud service providers. This method additionally evaluates and validates the data before performing the cryptographic operation. The encrypted data are protected from illegal access and change.

Cloud computing is used to provide universal, on-demand to a shared programmable asset [17]. The execution of a large volume of data handling is a dangerous problem for cloud storage services. The data size can be used to define each data copy. The cost of maintenance in viable cloud storage is reduced by adopting data-oriented solutions such as Dropbox, Mozy, and Memopal deduplication [18].

3.2.1 Dekey Approach

The Dekey method keeps track of convergent keys in a permanent manner. It gets rid of the remaining convergent keys and distributes them across several KM-CSPs. Dekey produces secret segments on the unique convergent keys and distributes these segments among several KM-CSPs as an alternative to encrypting convergent keys on a per-user basis. If a sequence of user segments is similar, a consistent convergent key can be used to access them. This effectively reduces the convergent keys' storage overhead. Dekey assigns acceptance responsibility to the convergent key and allows the convergent key to tolerate key mismatches in KM-CSP. Like the previous Dekey approach, the new Dekey approach effectively and persistently maintains the convergent keys. This method corrects a number of convergent keys generated during the deduplication process. These convergent keys are also dispersed over a number of KM-CSPs. For the chunk data, a unique convergent key will be generated as well.

3.2.2 Formulation of the Issue

The client, the cloud storage, and the key management for cloud service providers are the three units for outsourcing information to cloud storage. In order to reduce transfer transmission and storage capacity, the client must transfer the information to cloud storage. Other clients have access to the data, which is not uploaded to the cloud. It's used to store client information and then outsource that information to another client. Deduplication and cloud storage, which preserves unique data using convergent keys, are used to get rid of duplicate data. The data in an autonomous unit are represented using KM-CSPs. RSSS is used to disperse each convergent key over different KM-CSPs. Deduplication at the block level divides a document into smaller, fixed-size blocks, removing duplicate data blocks. For greater efficiency, these chunks are kept in cloud storage. Before being stored in the distributed storage, any duplicate data are labelled.

3.2.3 Threat Model and Security Objectives

In the thread management process, there are two categories of interlopers: external interlopers and private interlopers. Through an uncontrolled network, external intruders can have a certain familiarity with the data copy of nosiness. It takes on the role of a user interacting with the S-CSP. Interlopers of this type include competitors who use the S-CSP as a grateful distribution network. The S-CSP or any of the KM-CSPs could be referred to as private interlopers because they are trustworthy but obtrusive.

3.2.4 Blowfish

Blowfish is a block cypher with a hidden key. It stresses a 16-time encryption capability. The key can be up to 448-bit long and has a block size of 64 bits. Despite the fact that a complex startup stage is required before any encryption can take place. On large microchips, true data encryption is tremendously productive. Each round r consists of four actions, which are XORed together with the rth P-array entry to produce the left half of the data.

3.2.5 Lump Data Encryption and Decryption

Dekey has been implemented and used to deal with convergent keys used in the encryption and decryption processes. The block diagram of the fundamental model of this method is shown in Figure 2.1. For simplification, the file transfer and duplication module are employed. By executing different cores in parallel with the pipeline technique, the entire multicore features are utilized with respect to the existing CPU (Figure 3.15).

The private encryption was carried out with the explicit purpose of securing the hash keys while maintaining the master key. For each block, the SHA-256 method is used to generate 32-byte hash keys. The data are encrypted using the Blowfish process, and the hash keys are produced using the SHA-256 algorithm [19]. The cypher text of data is encrypted using the RC5 symmetric key encryption technique. The SHA-256 method generates the cypher text of the encrypted hash keys using the Blowfish encryption standard. The RSSS method is used to compare and implement the RC5, Blowfish, and SHA-256 algorithms [20].

3.2.6 Discussion and Conclusions

Clients use a shared key as a secret with multiple parameters such as security parameter, secret key, plaintext, and cypher text in symmetric key encryption. In the data deduplication process, the symmetric key encryption process is compared to the Dekey approach in Figure 3.16.

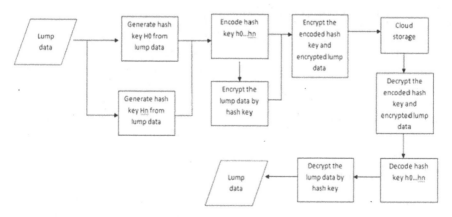

Figure 3.15 Encryption and decryption of lump data.

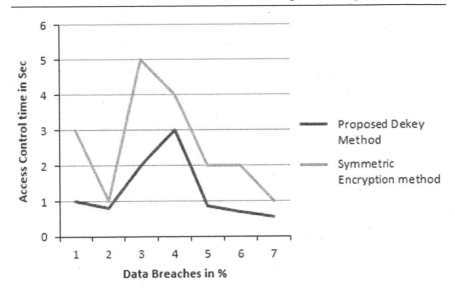

Figure 3.16 Comparison of the proposed Dekey method with the symmetric encryption method.

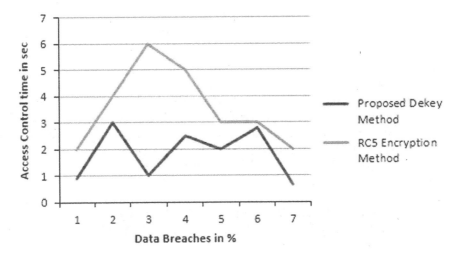

Figure 3.17 Comparison of the proposed method with the RC5 encryption method.

The convergent encryption process gives data dependability. Original data, convergent key, encryption procedure, and decryption process are all parameters used in this algorithm. Figure 3.17 shows a comparison of the convergent key and the suggested Dekey procedure. The data are encrypted using the RC5 algorithm in the form of a fixed-size word.

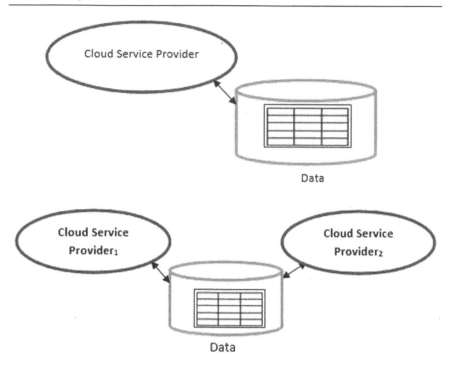

Figure 3.18 Comparison of proposed method with Blowfish method.

The w-bit plaintext, number of rounds, cypher text, encryption, and decryption processes are all parameters for this method. In Figure 3.17, the RC5 encryption algorithm is compared to the suggested Dekey approach. The Blowfish algorithm encrypts data using S-boxes with multiple rounds and an XOR operation. It includes security features such as block size, key array, function, encryption, and decryption. a comparison of Blowfish and the proposed Dekey method is shown in Figure 3.18.

3.3 DEDUPLICATION OF DATA WITH INTEGRATED PRIVACY TPA OVER CLOUD STORAGE METHOD

This section focuses on the data to maintain a TPA in cloud storage that is provided as a service. Only a certain standard of deduplication method is used by a single-level provider. To accomplish an integrated form of privacy preservation, different providers are used. This strategy employs a TPA that continuously monitors data during provider access. For the correct maintenance of the data within the boundary, data privacy regions are divided. For a better understanding of the cloud border, various lemmas are used. While transferring data from the client to the provider, various latencies are compared.

3.3.1 Formulation of the Problem

Normally, data are stored in a cloud storage with a variety of properties. The cloud plays a crucial role in data maintenance. The data are stored in the cloud, resulting in performance issues due to extra storage space and inconsistencies during the data handling process [21]. The deduplication is carried out in a single place, namely a single cloud vendor. Due to the same data being available in several cloud locations, this type of deduplication operation suffers from performance issues [22].

3.3.2 Work to Be Done

The data's privacy is carefully protected in order to prevent unintentional disclosure. There are two types of cloud vendors: non-malevolent and malicious [23]. The non-malicious vendor has access to all types of cloud services, whereas the malicious vendor does not have access. This challenge is solved by employing a high-security privacy preserved deduplication approach. The data are stored in the cloud using a deduplication approach that is only available from a single provider, as shown in Figure 3.19.

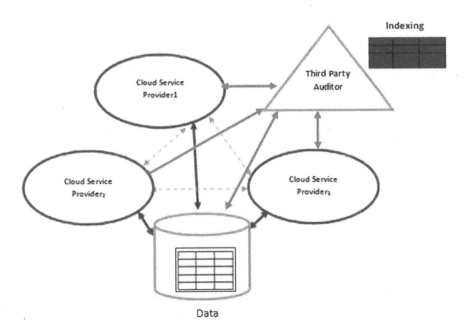

Figure 3.19 Single provider with deduplication process.

The issue is that the identical data may be available from another vendor, resulting in inconsistency. This difficulty is solved by utilizing many providers who execute data deduplication in a single location, as shown in Figure 3.18.

3.3.3 Proposed Data Deduplication with Integrated Privacy Protection

The proposed method employs a TPA in conjunction with an agreement to identify the malicious supplier for interaction. For service sharing, the providers interact with one another. The information is kept in a secure location [24].

The objective behind this clearance is that data in a centralized place can only be accessed by authorized providers. Following the clearance process, the TPA maintains the Index table for provider identification and assigns access permissions to the provider in the event of new data being inserted, current data being removed, or data being updated. The proposed solutions' major goal is to save data in a single location without redundancy, in a more secure manner, and with sufficient privacy protection at all levels. Figure 3.19 depicts the suggested method's overall architecture. The data in cloud storage are handled by a single provider, which has its own deduplication standards and methodology, as shown in Figure 3.20. Different providers have access to the same data, which can be managed safely and effectively. This challenge is solved by utilizing provider collaboration with common data maintenance in an efficient manner, as shown in Figure 3.21.

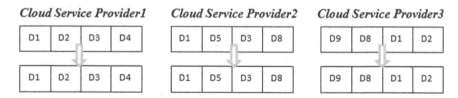

Figure 3.20 Multiple providers with deduplication process.

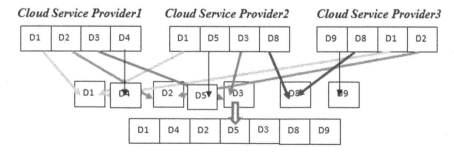

Figure 3.21 Proposed technique conceptual diagram.

3.3.4 Regions of Data Privacy

The practice of preventing unauthorized disclosure of data to a malevolent user is known as privacy maintenance. Customer level, TPA level, Provider level, and storage level are the four tiers of privacy boundaries. The data are kept private because it will go across multiple borders, so maintaining privacy is crucial. From the client level to the data storage level, there are many levels of data leakage.

3.3.5 Various Algorithms Are Compared

The deduplication process has a number of qualities that cloud storage supports in order to have a quick reaction time with no lag. From the client to the storage end, there are three types of latencies to consider: read latency, write latency, and communication delay. Storage gains improved throughput when latency over data or files in the cloud is kept to a minimum. Another attribute that gives efficient support during data transformation is the compression ratio. Without any redundant data, there are two approaches to analyze the compression ratio in customer and storage. The proposed method is used to compare the compression ratios of several algorithms such as DARE, iDeDup, and randomized convergent encryption (RCE). Over cloud storage, the deduplication rate of two layers is examined. DARE is a deduplication technique that detects data similarity and has a low overhead [25].

The identification ratio of similarity is 0.890%. iDeDup is a method for speeding up the deduplication process by reducing the amount of

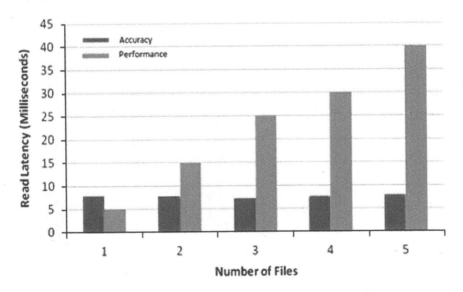

Figure 3.22 Single provider with data deduplication.

IO interaction and access time [26]. Latency is estimated to be 13% on average. RCE is a method for conducting deduplication in a more secure way [27]. The average delay percentage is 16%. Figures 3.22–3.24 demonstrate latency comparisons for various processes.

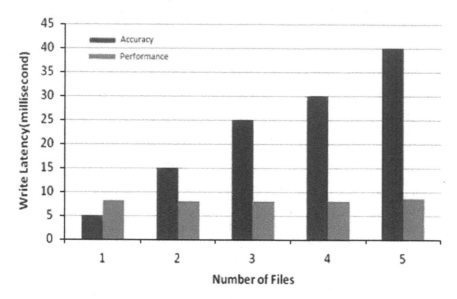

Figure 3.23 Data deduplication with multiple providers.

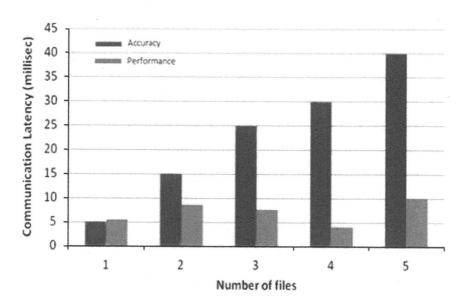

Figure 3.24 Comparison of read latency.

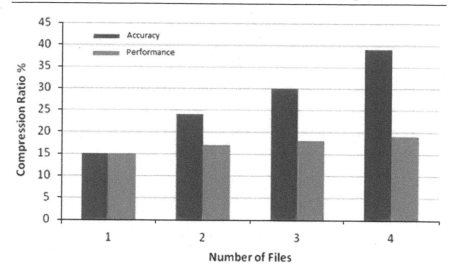

Figure 3.25 Comparison of write latency.

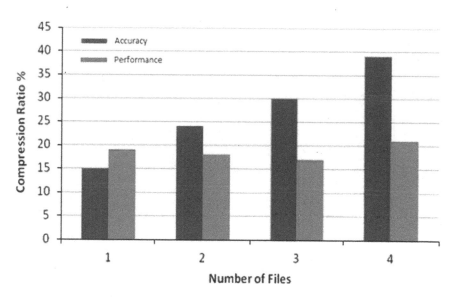

Figure 3.26 Comparison of communication latency.

Figure 3.25 shows the compression ratio of the deduplication procedure with different files at the customer end. Figure 3.26 depicts the cloud storage compression ratio. In Figure 3.27, various techniques linked to the deduplication process with compression ratio are illustrated.

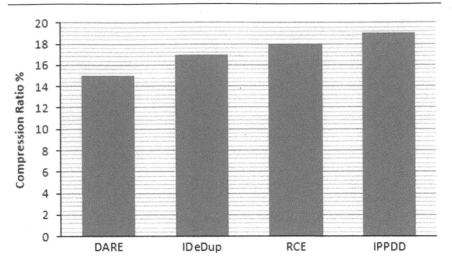

Figure 3.27 Client side compression radio.

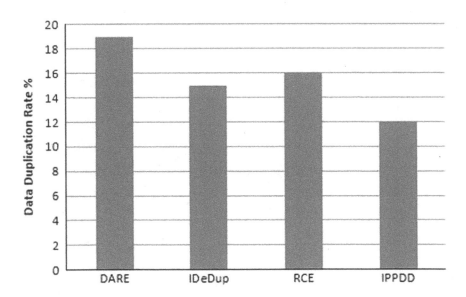

Figure 3.28 Compression ratio of cloud storage.

Figures 3.28 and 3.29 show the data deduplication rate of a single provider and an integrated provider, respectively, with varied parameters. When compared to existing techniques, the graph analysis shows that the suggested work delivers an efficient deduplication rate (Figures 3.30 and 3.31).

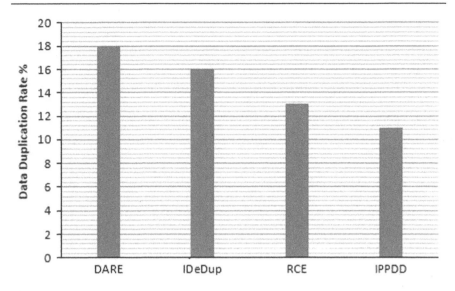

Figure 3.29 Integrated compression ratio.

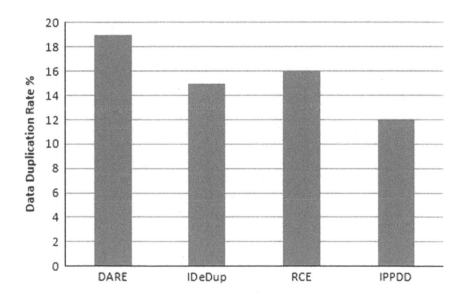

Figure 3.30 Single provider data deduplication rate.

3.4 CONCLUSION

The Dekey technique was used to secure client data in accordance with various cloud security standards. This approach secures bulky data, i.e., confidential data, by performing encryption and decryption processes. When uploading data to cloud storage, the hash-based technique is utilized

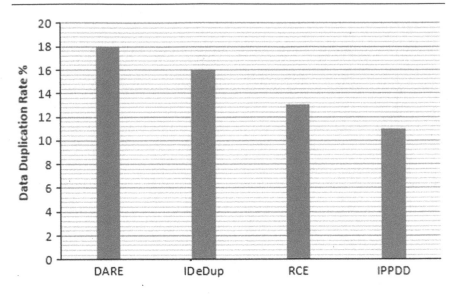

Figure 3.31 Integrated data deduplication rate.

to protect data integrity. The consumer must first decrypt the data before downloading it from the cloud storage and then confirm it for accuracy. The security risks associated with the deduplication process are detected and dealt with appropriately. With high reliability, various security algorithms are compared to the suggested Dekey technique.

For optimal reliability, an integrated privacy preservation of client data has been carried out. Along with the multi-level areas, a single-level provider with privacy regions is examined. After identifying the issues, the integrated privacy handling approaches are shown. The deduplication procedure is also evaluated by optimizing distinct definitions. Customer-level, third-party-level, provider-level, and storage-level privacy boundaries are all set. Various algorithms are compared in terms of read, write, and communication latency. Between single provider and integrated levels, the compression ratio is compared.

REFERENCES

[1] Baranwal, G., Vidyarthi, D. P., Baranwal, G., & Vidyarthi, D. P. (2015). A fair multi-attribute combinatorial double auction model for resource allocation in cloud computing. *The Journal of Systems & Software, 108*(no. Complete), 60–76. 10.1016/j.jss.2015.06.025

[2] Barba-Jimenez, C., Ramirez-Velarde, R., Tchernykh, A., Rodríguez-Dagnino, R., Nolazco-Flores, J., & Perez-Cazares, R. (2016). Cloud based video-on-demand service model ensuring quality of service and scalability. *Journal of Network and Computer Applications, 70*, 102–113. 10.1016/j.jnca.2016.05.007

[3] Beaumont, O., Eyraud-Dubois, L., & Lorenzo-Del-Castillo, J. A. (2014). Analyzing real cluster data for formulating allocation algorithms in cloud platforms. *Proceedings – Symposium on Computer Architecture and High Performance Computing*, 302–309. 10.1109/SBAC-PAD.2014.44

[4] Beltrán, M. (Nov. 2016). BECloud. *Future Generation Computing System*, 64(no. C), 39–49. 10.1016/j.future.2016.05.014

[5] Gulmezoglu, B., Inci, M. S., Irazoqui, G., Eisenbarth, T., & Sunar, B. (2016). Cross-VM cache attacks on AES. *IEEE Transactions on Multi-Scale Computing Systems*, 2(3), 211–222. 10.1109/TMSCS.2016.2550438

[6] Galante, G., & De Bona, L. C. E. (2012). A survey on cloud computing elasticity. *2012 IEEE Fifth International Conference on Utility and Cloud Computing*, 263–270.

[7] Wu, H., Wang, C., Lu, K., Fu, Y., & Zhu, L. (2018). One size does not fit all: The case for chunking configuration in backup deduplication. *Proceedings - 18th IEEE/ACM International Symposium on Cluster, Cloud and Grid Computing, CCGRID 2018*, 213–222. 10.1109/CCGRID.2018.00036

[8] Krishnaprasad, P. K., & Narayamparambil, B. A. (2013). A proposal for improving data deduplication with dual side fixed size chunking algorithm. *2013 Third International Conference on Advances in Computing and Communications*, 13–16. 10.1109/ICACC.2013.10

[9] Ma, K., Dong, F., & Yang, B. (2015). Large-scale schema-free data deduplication approach with adaptive sliding window using MapReduce. *The Computer Journal*, 58(11), 3187–3201. 10.1093/comjnl/bxv052

[10] Kundu, A., et al. (2010). Memory utilization in cloud computing using transparency. *5th International Conference on Computer Sciences and Convergence Information Technology*. 22–27. 10.1109/ICCIT.2010.5711023

[11] Coutinho, R. D. C., Drummond, L. M. A., Frota, Y., & De Oliveira, D. (2015). Optimizing virtual machine allocation for parallel scientific workflows in federated clouds. *Future Generation Computer Systems*, 46, 51–68. 10.1016/j.future.2014.10.009

[12] Lin, C., Cao, Q., Huang, J., Yao, J., Li, X., & Xie, C. (2018). HPDV: A Highly parallel deduplication cluster for virtual machine images, *Proceedings of the 18th IEEE/ACM International Symposium on Cluster, Cloud and Grid Computing*, 472–481. 10.1109/CCGRID.2018.00074

[13] Bellare, M., Keelveedhi, S., & Ristenpart, T., Message-locked encryption and secure deduplication. *Advances in Cryptology – EUROCRYPT 2013*, 2013, 296–312.

[14] Mao, B., Jiang, H., Wu, S., & Tian, L. (2016). Leveraging data deduplication to improve the performance of primary storage systems in the cloud. *IEEE Transactions on Computers*, 65(6), 1775–1788. 10.1109/TC.2015.2455979

[15] Fan, C.-I., Huang, S.-Y., & Hsu, W.-C. (2012). Hybrid data deduplication in cloud environment. *2012 International Conference on Information Security and Intelligent Control*, 174–177. 10.1109/ISIC.2012.6449734

[16] Su, H., Zheng, D., & Zhang, Y. (2017). An efficient and secure deduplication scheme based on rabin fingerprinting in cloud storage. *22017 IEEE International Conference on Computational Science and Engineering (CSE) and IEEE International Conference on Embedded and Ubiquitous Computing (EUC)*, 01, 833–836.

[17] Mell, P., & Grance, T. (2011). *The NIST definition of cloud computing*. Gaithersburg, MD: Special Publication (NIST SP), National Institute of Standards and Technology. 10.6028/NIST.SP.800-145

[18] Li, M. (2012). On the confidentiality of information dispersal algorithms and their erasure codes, 1–4 [Online]. Available: http://arxiv.org/abs/1206.4123.

[19] Shwetha, G. S. (2015). A secure data storage scheme in cloud using deduplication. *International Journal of Engineering Research & Technology*, *3*(27), 1–5.

[20] Verma, H. K., & Singh, R. K. (2012). Performance analysis of RC5, blowfish and DES block cipher algorithms. *International Journal of Computer Applications*, *42*(16), 975–8887. [Online]. Available: https://pdfs.semanticscholar.org/cdb7/b397e365338f7ddf78110c7ef7e190afca0e.pdf

[21] Hashizume, K., Rosado, D. G., Fernández-Medina, E., & Fernandez, E. B. (2013). An analysis of security issues for cloud computing. *Journal of Internet Services and Applications*, *4*(1), 5. 10.1186/1869-0238-4-5

[22] Leesakul, W., Townend, P., & Xu, J. (2014). Dynamic data deduplication in cloud storage. *2014 IEEE 8th International Symposium on Service Oriented System Engineering*, 320–325. 10.1109/SOSE.2014.46

[23] Tripathi, A., & Mishra, A. (2011). Cloud computing security considerations. *2011 IEEE International Conference on Signal Processing, Communications and Computing, ICSPCC, 2011* (no. October), 1–12. 10.1109/ICSPCC.2011.6061557

[24] More, S., & Chaudhari, S. (2016). Third party public auditing scheme for cloud storage. *Procedia Computer Science*, *79*, 69–76. 10.1016/j.procs.2016.03.010

[25] Rajagopalan, A. X. B. F. S., & Ramamoorthy, S. (2018). Integrated privacy preserving data deduplication method using third party auditor over cloud storage. International Journal of Engineering Research & Technology. 7, 422–426.

[26] {iDedup}: Latency-aware, Inline Data Deduplication for Primary Storage. (Feb. 2012), [Online]. Available: https://www.usenix.org/conference/fast12/idedup-latency-aware-inline-data-deduplication-primary-storage.

[27] Shynu, P. G., Nadesh, R. K., Menon, V. G., Abbasi, V. P. M., & Khosravi, M. R. (2020). A secure data deduplication system for integrated cloud-edge networks. *Journal of Cloud Computing*, *9*(1), 61. 10.1186/s13677-020-00214-6

Chapter 4

Internet of Things for Smart Cities: Application and Challenges

Arun Kumar Rana, Priya Trivedi, and Paras Chawla

CONTENTS

4.1 INTRODUCTION

The phrase "smart city" refers to new sectors that make use of information and communication technologies (ICT), as well as urban functions and environments [1]. The word refers to the integration of ICT with combination and urban operations in a restricted sense. The term smart city, on the other hand, can be defined in a broad sense as the confluence of ICT, energy technology, the environment, and support facilities in residential environments and urban [2,3]. Various sensors, support technologies, and backdrop settings are required to prepare the fundamental infrastructure of a smart city, and they are being used in urban areas. The Internet of Things (IoT) is regarded as one of the most critical factors for the effective deployment of a smart city [4]. "A set of technologies for accessing the data collected by various devices through wireless and wired Internet networks," according to the IoT [5]. Although there are significant discrepancies in definitions of IoT, a popular explanation is the potential of diverse user devices to deliver important and beneficial information over wireless and wired Internet networks. A big number of people have recently changed their lifestyles and moved to metropolitan areas. Around 70% of the population is expected to be reliant on the urban environment. As a result, some of the systems with sensors are employed in IoT devices for smart city development. The smart city concept proposes a novel approach to developing additional gadgets in the fields of transportation, healthcare, housing, buildings, and the environment in which we live. The IoT is being referred to as a new network accessing technology. The term "Internet of Things" simply refers to a thing or device that is connected to the internet in order to collect and exchange massive amounts of data.

Things refer to a variety of objects and gadgets that are connected to the Internet. Several types of sensors, such as software, wireless sensors, and other computer equipment, are used in IoT devices. These devices assist in the transfer and sharing of relevant data over the internet without the need for human-to-human or human-to-computer interaction, as well as information about the surroundings. Because the IoT and smart cities are relatively new concepts, there isn't yet a huge corpus of literature on the issue. As research in the sector evolved and new technologies matured, a number of businesses, including CISCO, IBM, and others, produced solutions that made IoT a viable alternative for modern cities looking to improve the quality of life.

A new buzzword, the Internet of Things, is now having an impact on our lives.

Globalization has progressed, and the world is now genuinely borderless, as individuals are increasingly connected not only through technology but also through social media. The Internet network connects items or electronic devices, allowing for the creation of new services that are both innovative and beneficial. IoT refers to a system in which the Internet is linked to the physical world through a network of sensors. By 2011, the number of Internet-connected devices on the earth [12.5 billion] has overtaken the number of people on the planet [7 billion]. By 2020, the number of Internet-connected devices is estimated to be between 26 billion and 50 billion worldwide.

This chapter discusses the role of IoT technology in the creation of a smart city. The following is how the rest of the chapter is organized: Section 4.2 gives an overview of the many IoT technologies employed by various scientists. Section 4.3 discusses the benefits of IoT in smart cities. Section 4.4 discusses the challenges of implementing IoT in smart cities. Finally, Section 4.5 summarizes this study's findings [6–13].

4.2 LITERATURE REVIEW

Since the development of various forms of networks, IoT has become one of the most important types of infrastructure in smart cities.

In a smart home setting, for example, data collected by electronic household appliances such as refrigerators are shared and saved to enable user-customized services [13–17]. Similar to the concept of a smart home environment, the smart city is a growing market and a critical part of future infrastructure. In a smart city, the importance of IoT technology is magnified since it aims to use energy and power efficiently, resulting in a convenient and economically sound infrastructure for society's well-being. As a result, once a smart city's main infrastructure is in place, many services that need the utilization of various types of data obtained in daily life can be provided. This means that a smart city's myriad IoT-enabled services can assist residents in living in a more sustainable and joyful environment. As a result, a number of important industries are collaborating to develop a more sustainable urban environment, taking into account the technological and social challenges that future smart city concepts would entail. In addition, earlier research summarized the key technological features of smart city technologies, and IoT technologies can be considered a pre-requisite based on this review [14,18–24].

4.3 IOT BENEFITS IN SMART CITIES

IoT is used by businesses for creative management and remote monitoring of processes. As a result, they can even operate the latter from afar because data are constantly fed into applications and data storage. The IoT gives you the advantage of knowing what's going on ahead of time. Because of the low cost of IoT, hitherto unreachable activities may now be monitored and managed. The most significant benefit is financial because this new technology has the potential to replace humans in charge of monitoring and maintaining supply. As a result, costs can be cut and streamlined greatly. IoT also allows for completely new insights, such as the correlation of meteorological effects with industrial production [25–33].

4.3.1 Smart Infrastructure

Cities must create the conditions for long-term development: digital technologies are becoming more vital, and urban infrastructure and buildings must be developed more efficiently and sustainably. CO_2 emissions should be kept to a minimum, such as by investing in electric cars and self-propelled vehicles. Smart lights should only turn on when someone walks past them, such as establishing brightness levels and recording daily use to reduce the demand for electrical power in smart cities. Figure 4.1 shows smart infrastructure.

Figure 4.1 Smart infrastructure.

4.3.2 The City Air Management Tool

Siemens has developed "The City Air Management Tool (CyAM)," a comprehensive cloud-based software suite with the following features: By recording pollution data in real-time and forecasting emissions, it can gain emissions for the next three to five days with up to 90% accuracy. CyAM is a cloud-based software suite that includes a dashboard that shows real-time data on air quality discovered by sensors throughout a city and forecasts values for the next three to five days. To simulate the next three to five days, cities can choose from 17 different measures (effects of the air quality for the upcoming three to five days). Consequences: All of these options include new environmental zones (low-emission zones), speed limits, and free public transportation. CyAM is based on Mind Sphere, Siemens' cloud-based, open IoT operating system [34–37]. Figure 4.2 shows CyAM.

4.3.3 The Challenge for Huge Smart Cities Is to Optimize Traffic Flow

Los Angeles: As one of the world's busiest cities, Los Angeles has developed an intelligent transportation technology to manage traffic flow. Traffic flow sensors embedded in the pavement give real-time updates to a central signal control platform, which analyses the data and adapts traffic lights to the current traffic condition in seconds. It predicts where traffic will travel based on historical data — all without the need for human intervention. Figure 4.3 shows IoT for smart cities' traffic flow.

Figure 4.2 City air management tool (CyAM).

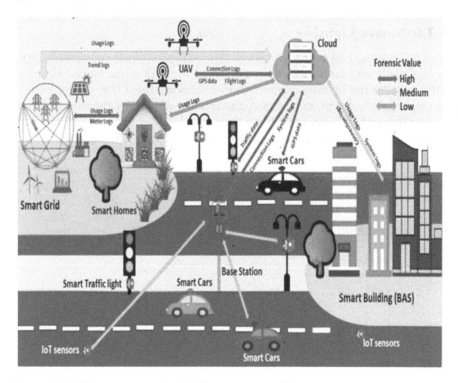

Figure 4.3 IoT for smart cities' traffic flow.

4.3.4 Smart Parking

The sensors on the ground report to the motorist through their smartphone where they may find a free parking place. Intelligent parking solutions determine when a car has left the parking area. Others employ vehicle input to pinpoint exact locations of openings and urge waiting automobiles into the path of least resistance. Smart parking is a reality today, requiring no complicated infrastructure or significant expenditure, making it appropriate for a mid-sized smart city.

4.3.5 Smart Waste Management

Waste management solutions aid in improving waste collection efficiency, lowering operational costs, and addressing environmental challenges connected with inefficient waste collection. A level sensor is installed in the waste container, and when a particular threshold is achieved, a notification is sent to the truck driver's management platform via smartphone. The message appears to empty a full container, avoiding drains that are half full [38–40]. Figure 4.4 shows IoT for smart waste management.

4.3.6 Smart Lighting

Hue & Lighting Holding B.V.'s smart illumination systems can sense human beings easily and automatically detect when they are in the room, allowing them to alter the lighting as needed. Furthermore, there is a gadget known as smart light bulbs that can detect and modify or regulate themselves based on the amount of light available during the day. Remote technologies can be used to control it. Figure 4.5 shows IoT with smart lighting.

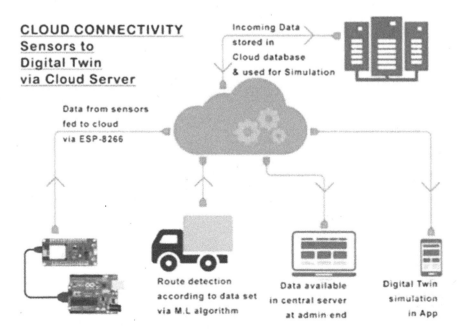

Figure 4.4 IoT for smart waste management.

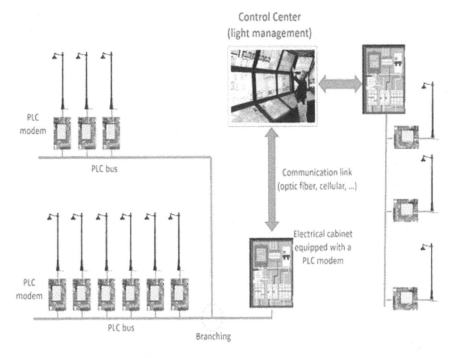

Figure 4.5 IoT with smart lighting.

4.3.7 Crowd Sensing

Crowd sensing is a powerful technique that enables better monitoring of the environment and control the air pollution in congested areas. Users can analyze the air quality in their immediate vicinity using IoT devices and technology with this crowd sensing approach. It's utilized in smart cities to give objects or smart devices known as sensors a high level of intelligence, allowing them to integrate and coordinate better. By allowing the environment to connect with each gadget and detecting their varied behaviours, the smart city is produced. Allow these smart devices to collect data and generate more precise results based on environmental concerns. One of the most essential IoT application systems in smart cities and the environment is crowd monitoring. Various device and sensor technologies are actively involved or contribute to the collection of accurate data about the huge number of persons involved in the process. When all of the data has been gathered, it is routed to a central server, which stores it all and uses it for following procedures. In order to improve their lifestyle, the central server will perform a poll and provide feedback to residents entirely through responses and activities. By giving an alternative source to air quality stations, the crowd sensing device system is supplied. Only the tiniest sensors, which are assigned to a huge group of individuals, are used in traditional monitoring stations, and the system works properly. Figure 4.6 shows architecture of crowd sensing system

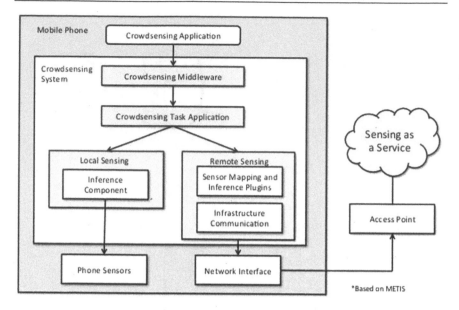

Figure 4.6 Architecture of crowd sensing system [7].

4.4 CHALLENGES FOR IMPLEMENTATION OF IOT IN SMART CITY

The implementation of IoT is dependent on a number of important aspects. Understanding the implementation problems may aid management in the design, direction, and control of smart city services. In order to deliver better quality services, policymakers might plan for IoT implementation. Health concerns traffic congestion, insufficient power, inadequate housing, environmental degradation, and an increase in crime rates are all issues that have arisen as a result of the city's rapid growth. The IoT is seen as the next big thing in the world of technology. As a computing device, the IoT connects various electronic appliances and devices, interprets and understands human interaction, and achieves high-quality two-way communication. However, putting IoT technologies in place to achieve urban sustainability is a multidimensional process. The outcomes of this study suggest that IoT device interoperability is a major issue, as the interoperability challenge is to make all IoT devices work together in integrated software platforms. Companies are designing IoT solutions on their own, based on their own needs, using multiple platforms, resulting in poor device integration, which is currently only achievable through pre-programmed APIs. To promote an interoperable framework, companies and IoT-based organizations must establish open-source platforms. This will make the integration of any hardware or software across all platforms much easier. The level of talent necessary to operate IoT devices, which run autonomously, is the next great hurdle. This implies that users should be able to inspect the hardware and software development of their devices. Furthermore, users must be capable of visualizing, interpreting, and analyzing advanced data. Because IoT systems in a smart city setting must be able to deal with mobile data sources, city infrastructure mobility is also a critical issue. IoT devices from various manufacturers obstruct seamless data connection and exchange since they gather data in different forms and run on different operating systems.

The lack of seamless collaboration stymies the smartification of services, resulting in an island of non-collaborative and non-standardized services. Due to a lack of standards, IoT devices are poorly built, which has a severe influence on network use. Standardization of IoT devices, on the other hand, will lead to interchangeability and, as a result, greater utilization of their resources. In a smart city with IoT devices, security and privacy are critical concerns. Data containing private/confidential information about citizen behaviour and government policies must be safeguarded against unauthorized access.

High-security procedures must be implemented across the network by agencies in charge of supplying a variety of smart services. IoT devices exchange and communicate data regarding a citizen's health-related medical records along with financial records in a smart city environment. Citizens will be profiled as a result of high-profile cyberattacks and data breaches, infringing on their privacy. Figure 4.7 shows implementation of IoT in smart city.

Figure 4.7 Implementation of IoT in smart city.

4.5 CONCLUSION

In this chapter, we looked at how IoT can be used in smart cities and the environment to improve our lives. A new environment will emerge in the future as a result of the use of IoT technology, including cities with smart houses. The IoT is assisting our cities and environment in becoming smarter, according to this chapter. The benefits of utilizing sensors in smart homes and cities are discussed extensively. They also go into great detail on how they deal with environmental issues. Smart environment applications are more difficult to implement than other systems. However, once they are correctly applied, they can give significant benefits to our society. As a result, it aids in improved environmental development, and facilities and use become easier. It creates a good environment with their smart devices or sensors, as well as their IoT solutions. The IoT is the greatest option among all those that aid in the development of smart cities and the environment. The performance of IoT technology is improved by addressing the benefits of the technology in smart cities and the environment. The IoT's potential is boundless. Healthcare, manufacturing, restaurants, fashion, and education are among the industries where it provides services. Smart cities can share a single platform, which is particularly beneficial for small towns. IoT solutions for smart cities are cloud-based, which is ideal given that they share a platform based on open data. Small towns can work together to create a common urban ecosystem. As a result, both small and large smart city solutions are connected and controlled by a single cloud platform. Finally, and perhaps most crucially, the size of a city is not a determining factor in its ability to become "smart," as cities of all sizes can benefit from intelligent technologies.

REFERENCES

[1] Duygan, M., Fischer, M., Pärli, R., & Ingold, K. (2022). Where do smart cities grow? The spatial and socio-economic configurations of smart city development. *Sustainable Cities and Society*, 77, 103578.

[2] Guo, M., Xia, M., & Chen, Q. (2022). A review of regional energy internet in smart city from the perspective of energy community. *Energy Reports*, 8, 161–182.

[3] Kumar, A., Sharma, S., Dhawan, S., Goyal, N., & Lata, S. (2021). 14 E-learning with. *Internet of Things: Robotic and Drone Technology*, 195.

[4] Dhawan, S., Chakraborty, C., Frnda, J., Gupta, R., Rana, A. K., & Pani, S. K. (2021). SSII: secured and high-quality steganography using intelligent hybrid optimization algorithms for IoT. *IEEE Access*, 9, 87563–87578.

[5] Li, X., Liu, H., Wang, W., Zheng, Y., Lv, H., & Lv, Z. (2022). Big data analysis of the internet of things in the digital twins of smart city based on deep learning. *Future Generation Computer Systems*, 128, 167–177.

[6] Kusumastuti, R. D., Nurmala, N., Rouli, J., & Herdiansyah, H. (2022). Analyzing the factors that influence the seeking and sharing of information on the smart city digital platform: empirical evidence from Indonesia. *Technology in Society*, 101876.

[7] Dhawan, S., Gupta, R., Rana, A., & Sharma, S. (2021). Various swarm optimization algorithms: Review, challenges, and opportunities. *Soft Computing for Intelligent Systems*, 291–301.

[8] Kumar, A., & Sharma, S. (2021). Internet of robotic things: design and develop the quality of service framework for the healthcare sector using CoAP. *IAES International Journal of Robotics and Automation*, 10(4), 289.

[9] Khan, M. A., & Salah, K. (2018). IoT security: review, blockchain solutions, and open challenges. *Future Generation Computer Systems*, 82, 395–411.

[10] Rana, A. K., Krishna, R., Dhwan, S., Sharma, S., & Gupta, R. (2019, October). Review on artificial intelligence with Internet of things – problems, challenges and opportunities. In *2019 2nd International Conference on Power Energy, Environment and Intelligent Control (PEEIC)* (pp. 383–387). IEEE.

[11] Rana, A. K., & Sharma, S. (2021). Contiki Cooja Security Solution (CCSS) with IPv6 routing protocol for low-power and lossy networks (RPL) in Internet of things applications. In *Mobile Radio Communications and 5G Networks* (pp. 251–259). Singapore: Springer.

[12] Rana, A. K., & Sharma, S., 2019. Enhanced energy-efficient heterogeneous routing protocols in WSNs for IoT application. *International Journal of Engineering and Advanced Technology*, 9, 4418–44250.

[13] Ahmed, E., Islam, A., Ashraf, M., Chowdhury, A. I., & Rahman, M. M. Internet of Things (IoT): vulnerabilities, security concerns and things to consider. *123*, 1110–1127.

[14] Veeramanickam, M. R. M. & Mohanapriya, M. (2016). IoT enabled futurus smart campus with effective e-learning: i-campus. *GSTF Journal of Engineering Technology (JET)*, 3(4), 8–87.

[15] Cho, S. P. & Kim, J. G. (2016). E-learning based on Internet of Things. *Advanced Science Letters*, 22(11), 3294–3298.

[16] Kumar, A., & Sharma, S. Demur and routing protocols with application in underwater wireless sensor networks for smart city. In *Energy-Efficient Underwater Wireless Communications and Networking* (pp. 262–278). IGI Global.

[17] Abbasy, M. B., & Quesada, E. V. (2017). Predictable influence of IoT (Internet of Things) in the higher education. *International Journal of Information and Education Technology*, 7(12), 914–920.

[18] Kumar, A., Salau, A. O., Gupta, S., & Paliwal, K. (2019). Recent trends in IoT and its requisition with IoT built engineering: a review. In *Advances in Signal Processing and Communication* (pp. 15–25). Singapore: Springer.

[19] Charmonman, S., Mongkhonvanit, P., Dieu, V., & Linden, N. (2015). Applications of Internet of Things in e-learning. *International Journal of the Computer, the Internet and Management*, 23(3), 1–4.

[20] Vharkute, M., & Wagh, S. (2015, April). An architectural approach of Internet of Things in E-Learning. In *2015 International Conference on Communications and Signal Processing (ICCSP)* (pp. 1773–1776). IEEE.

[21] Li, H., Ota, K., & Dong, M. (2018). Learning IoT in edge: deep learning for the Internet of Things with edge computing. *IEEE Network, 32*(1), 96–101.

[22] Dalal, P., Aggarwal, G., & Tejasvee, S. (2020). Internet of Things (IoT) in Healthcare System: IA3 (Idea, Architecture, Advantages and Applications). *Available at SSRN 3566282.*

[23] Rana, A. K., & Sharma, S. Industry 4.0 manufacturing based on IoT, cloud computing, and big data: manufacturing purpose scenario. In *Advances in Communication and Computational Technology* (pp. 1109–1119). Singapore: Springer.

[24] Wang, Q., Zhu, X., Ni, Y., Gu, L., & Zhu, H. (2020). Blockchain for the IoT and industrial IoT: a review. *Internet of Things, 10,* 100081.

[25] Rana, A. K., Salau, A., Gupta, S., & Arora, S., 2018. A survey of machine learning methods for IoT and their future applications. *23,* 110–121.

[26] Kumar, K., Gupta, E. S., & Rana, E. A. K. Wireless sensor networks: a review on "challenges and opportunities for the future world-LTE".*Amity Journal of Computational Sciences (AJCS), 1*(2).

[27] Sachdev, R. (2020, April). Towards Security and Privacy for Edge AI in IoT/IoE based Digital Marketing Environments. In *2020 Fifth International Conference on Fog and Mobile Edge Computing (FMEC)* (pp. 341–346). IEEE.

[28] Simic, K., Despotovic-Zrakic, M., Đuric, I., Milic, A., & Bogdanovic, N. (2015). A model of smart environment for e-learning based on crowdsourcing. *RUO. Revija za Univerzalno Odlicnost, 4*(1), A1.

[29] Kim, T., Cho, J. Y., & Lee, B. G. (2012, July). Evolution to smart learning in public education: a case study of Korean public education. In *IFIP WG 3.4 International Conference on Open and Social Technologies for Networked Learning* (pp. 170–178). Berlin, Heidelberg: Springer.

[30] An, S., Lee, E., & Lee, Y. (2013). A Comparative Study of E-Learning System for Smart Education. *International Association for Development of the Information Society.*

[31] El-Ala, N. A., Awad, W. A., & El-Bakry, H. (2012). Cloud computing for solving E-learning problems. *Editorial Preface, 3*(12).

[32] Rana, A., Chakraborty, C., Sharma, S., Dhawan, S., Pani, S. K., & Ashraf, I. (2022). Internet of medical things-based secure and energy-efficient framework for health care. *Big Data, 10*(1), 18–33.

[33] Rana, A. K., Sharma, S., Dhawan, S., & Tayal, S. (2021). Towards secure deployment on the internet of robotic things: architecture, applications, and challenges. In *Multimodal Biometric Systems* (pp. 135–148). CRC Press.

[34] Arora, S., Sharma, S., & Rana, A. K. (2021). 8 Ultrawide band antenna for wireless communications. *Internet of Things: Robotic and Drone Technology, 95.*

[35] Kumar, A., Sharma, S., Goyal, N., Singh, A., Cheng, X., & Singh, P. (2021). Secure and energy-efficient smart building architecture with emerging technology IoT. *Computer Communications, 176,* 207–217.

[36] Gupta, A., Kansal, A., & Chawla, P. (2021). Design of a wearable MIMO antenna deployed with an inverted U-shaped ground stub for diversity performance enhancement. *International Journal of Microwave and Wireless Technologies, 13*(1), 76–86.

[37] Eashwar, S., & Chawla, P. (2021, June). Evolution of agritech business 4.0 – architecture and future research directions. In *IOP Conference Series: Earth and Environmental Science* (Vol. 775, No. 1, p. 012011). IOP Publishing.

[38] Singh, S., & Chawla, P. (2021, May). Design and performance analysis of broadband Doherty power amplifier for 5G communication system. In *2021 2nd International Conference for Emerging Technology (INCET)* (pp. 1–4). IEEE.

[39] Jain, P., & Chawla, P. (2021, September). A novel smart healthcare system design for Internet of health Things. In *2021 International Conference on Innovative Computing, Intelligent Communication and Smart Electrical Systems (ICSES)* (pp. 1–8). IEEE.

[40] Kaur, H., & Chawla, P. (2022). Performance analysis of novel wearable textile antenna design for medical and wireless applications. *Wireless Personal Communications*, 1–17.

Chapter 5

IoT-Based Garbage Monitoring System with Proposed Machine Learning Model for Smart City

Arun Kumar Rana, Sachin Dhawan, Sumit Kumar Rana, and Sandeep Kajal

CONTENTS

5.1 INTRODUCTION

The Internet of Things (IoT) and other communication technologies have greatly improved our ability to comprehend our surroundings. The adoption of IoT technologies, which have the ability to gather and analyse data about the surrounding environment, may improve life quality [1]. This situation helps the development of smart cities by making it easier for things and people to communicate with one another. By the end of 2020, there were expected to be 50 billion IoT devices [2,3]. The IoT is a complex and interconnected system. As a result, meeting the security needs of an IoT system with a huge attack surface is difficult. The increased use of IoT has unintentionally resulted in IoT deployment being an integrated process. There are a lot of concerns to bear in mind while establishing an IoT system: security, energy efficiency, analytics methodologies, and interoperability with other software applications [3]. IoT devices, on the other hand, frequently function without the assistance of a human operator. As a result, an intruder can physically get access to these devices. Eavesdropping on wireless networks used by IoT devices, which are frequently connected by a communication channel, allows intruders to obtain access to sensitive information. The area, watching and the officials of wastes are a portion of

DOI: 10.1201/9781003355960-5

the fundamental issues of right now. A blundering method is a traditional strategy for truly testing the misfortunes of waste holders and eventually requires human initiative, time, and expense that can be eliminated with our latest advancements without much stretch [3–6]. In comparison to WiFi, MQTT makes the job all the more easy and fast. The pressure sensor detects the dry and wet presence and then, according to the dry and wet bin, it will divide the waste accordingly [7–11]. The ultrasonic sensor is used to recognize the garbage-filled state and update the garbage collection management on the bin condition accordingly [12,13]. As transport needs to clean whether the dustbin is filled or depleted, this chapter contributes to the exhaustion of time. It can also create additional traffic and noise. This method demands high costs and can produce an atmosphere that is unhygienic and can cause illnesses and health problems [1,2]. The IoT which interconnects human and social networks has a specific method and algorithm, it is not only about the human and social network but also the sociality of the device [19,20]. Machines are outfitted with IoT sensors that monitor discrete factors such as vibration, noise, heat, and temperature. After that, the data are transferred to the cloud for analysis. Now comes machine learning, with the machine learning model sitting on the cloud platform and feeding on incoming data. The ML model divides the data into two categories: training and verification. To generate a hypothesis, the model examines hundreds of thousands of records for anomalies, correlations, and projections. The hypothesis must be tested and validated after it has been established. A model is published as an executable endpoint once it has been validated.

5.2 IOT DATA PROTOCOLS

As shown in Figure 5.1 portion of the IoT information protocols are as follows:

a. **Message Queue Telemetry Transport (MQTT)**
   One of the most favoured protocols for IoT gadgets, MQTT gathers information from different electronic gadgets and supports remote gadget checking. It is a subscribe/publish convention that runs over transmission control protocol (TCP), which implies it bolsters occasion-driven message trade through remote systems [18].
b. **CoAP (Constrained Application Protocol)**
   CoAP is a web utility convention for IoT devices. Utilizing this convention, the customer can send a solicitation to the server and the server can send back the reaction to the customer in HTTP. For lightweight execution, it utilizes user datagram protocol (UDP) and decreases space utilization. The convention utilizes twofold information group efficient XML (EXL) interchanges. CoAP convention is utilized basically in

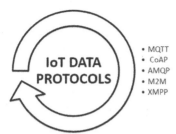

Figure 5.1 IoT data protocols.

robotization, mobiles, and microcontrollers. The convention sends a solicitation to the application endpoints, for example, machines at homes and sends back the reaction of administrations and assets in the application [21].

c. **Advanced Message Queuing Protocol (AMQP)**

AMQP is a software layer convention for message-arranged middleware condition that gives routing and queuing. It is utilized for solid point-to-point associations and supports the consistent and secure trade of information between the associated gadgets and the cloud. AMQP comprises three separate parts in particular Exchange, Message Queue, and Binding.

d. Machine-to-Machine (M2M) Communication Protocol

It is an open industry protocol that worked to give the remote application the executives of IoT gadgets. M2M correspondence conventions are financially savvy and utilize open systems. It makes a situation where two machines impart and trade information.

e. Extensible Messaging and Presence Protocol (XMPP)

The XMPP is particularly structured. It utilizes a pushing instrument to trade messages progressively. XMPP is adaptable and can incorporate the progressions flawlessly. Created utilizing open XML (Extensible Markup Language), XMPP fills in as a nearness marker demonstrating the accessibility status of the servers or gadgets transmitting or getting messages.

5.3 MQTT PROTOCOL WORKING AND USES

IoT allows the association of web-based gadgets with the ability to assemble and trade data. Typically, these gadgets are added with smaller scale controllers such as Arduino, sensors, actuators, and web accessibility. In this unique situation, without knowing each other's personalities, MQTT assumes a significant job of trading information or data between IoT gadgets. This chapter introduces distinctive IoT interaction assistance models [22–25]. Model A presents the use of sequential USB as a communication medium, while Model B uses the MQTT to connect the system

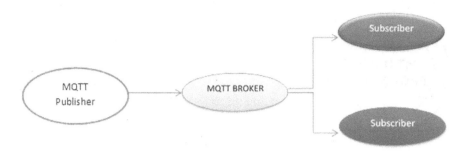

Figure 5.2 MQTT information flow.

with the internet by sending a WiFi cable (ESP8266-12). The definition of distributors and endorsers is used for communications, as seen in Figure 5.2. Through the help of a merchant or computer, messages are transmitted or acquired. This operator is responsible for scattering messages that depend on the judgement of the topic of a message to schedule clients. Additionally, the processor is called the merchant in MQTT. Mosquitto, Adafruit, and hiveMQ are a few agents used in MQTT. MQTT is a lightweight production/membership type (bar/sub) informing convention, probably the most widely acquired standard in the Industrial IoT to date. MQTT's tech, intended for battery-fuelled devices, is simple and lightweight, granting devices low use of force. Dealing with the top of the TCP/IP protocol, it was uniquely intended for dubious correspondence organisations to respond to the problem of the increasing number of small low-power protests that have shown up in the framework in the current years, little calculated [14–16].

MQTT relies upon the model of the subscriber, editor, and broker. The publisher's business is to collect the information within the model and send data to the subscriber using the intervention layer that is the broker. Due to the reality of sending messages over TCP and monitoring long theme names, MQTT can be dangerous for certain prohibitive devices, considering its characteristics. With the MQTT-SN variation that utilizes UDP and supports the order of the subject name, this is unravelled. Nevertheless, despite its broad range, MQTT does not endorse the board structure model with a very characterized information representation and gadget, which makes the execution of its information fully transparent by the board and gadget of the capability of the executive or seller [17,18].

For holes or vandalism, MQTT empowers applications such as watching a gigantic oil pipeline. Those must be amassed into a sole field for examination by a large amount of sensors. It will make a step to fix the problem at the stage where the structure detects a problem. Multiple MQTT uses include force usage research, lighting management, and even logical cultivation. They express an obligation to collect and make it available to the IT foundation data from various outlets.

5.4 WORKING FLOW OF MQTT-BASED SYSTEM

The enormous measure of the earth's population (i.e., 70%) would migrate to metropolitan centres by 2050, thereby creating colossal cities [1]. These urban areas need a genius economic base for coping with the needs of people and providing programmes that are essential and further developed [2]. The idea struck us when we saw that, as seen in Figure 5.3, the dump truck used to bypass the city to collect heavy waste twice a day. Despite the strength of this system, it was inefficient. For example, imagine road A is going to be a busy road and we see that the garbage tops off very quickly, while the container is not even half full after two days, maybe road B dramatically.

What our system does is that it provides at any random moment an on-going pointer of the trash stage in a trashcan. Using that data, we will then be able to enhance squander variety of courses and gradually decrease fuel usage. It helps garbage authorities to design their day to achieve a plan by day/week after week.

The algorithm for the flow chart is as follows (Figure 5.4):

Stage 1. Start
Stage 2. Check all components of the framework are working accurately or not
Stage 3. If any segment not working or battery is less or the top isn't shut then it will send a warning
Stage 4. If everything is right then the status will be online to the server else it'll be disconnected
Stage 5. Module will begin the ultrasonic sensor to check the fill levels of container
Stage 6. Values of the ultrasonic sensor are utilized to compute: Distance = Speed*Time
Stage 7. Module at that point sends fill level to the server after each 1 moment

Figure 5.3 Different bins with sensor.

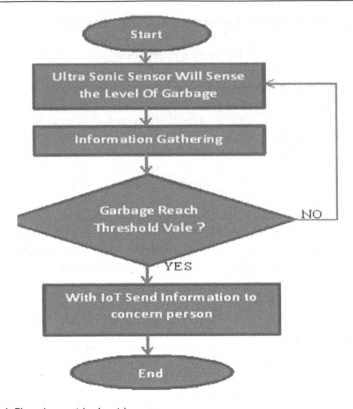

Figure 5.4 Flowchart with algorithms.

Stage 8. Go to Step 6
Stage 9. End

5.5 MQTT-BASED SYSTEM WITH RESULT

In the present system, waste is collected in two or three days and due to this waste overflows and creates huge pollution by spreading bad smells and diseases. The existing system only carries the Arduino on with a WiFi device which is very costly and not much effective for the present system but our proposed system is very useful to the people and garbage management system. This helps to keep the city clean and helps in the vision of Smart City. This works if the percentage of garbage in a bin is greater than 75% then a notification will be sent to the vehicle to clean the garbage bin. The percentage also can be seen by the authorities with the help of a database and webpage developed. Right now, squander is overseen in two distinctive wastes likewise, for example, dry and wet. MQTT is a very lightweight protocol that consumes very little RAM in the size of kilobytes.

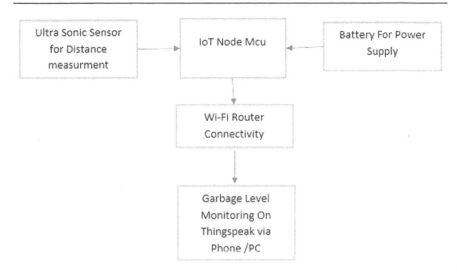

Figure 5.5 Proposed systems.

The proposed framework effectively screens the degree of the trashcan utilizing the ultrasonic sensor as indicated the Figure 5.5.

The NodeMCU unit is utilized to interface all the data gathered from the ultrasonic sensor and the website page. Each ESP8266 module comes pre-changed with an AT request set firmware. The ESP8266 module is an incredibly monetarily astute board with a huge, and ever-creating, arrangement. Ultranationalistic sensors measure that unit of the goal toward evaluating the long stretch between those overflowing and assembling. That trigger could a chance to be seen as a fundamental pulse that turns the sensor on, every the long run a detachment will be evaluated. It should last 10 smaller scale seconds.

5.6 WORKING AND HOW TO USE THINGSPEAK

Matlab provides an amazing online service for the IoT project in the form of a cloud and the famous cloud is ThingSpeak, and to use this cloud – first, you have to create an account in this cloud and with that, it is free for every user and then you have to click on channel mentioning the things that take the website in the channel setting and gives your project name and introduces some main topic and description of the project in the given field. Then you have to activate this field to show the level of garbage. Before activation you have to copy the API key from your channel then you have to paste it into the program so that communication between hardware and cloud establish as shown in Figure 5.6. The framework utilizes an ultrasonic sensor to put over the containers to identify the trash level and

Figure 5.6 ThingSpeak output result.

contrast it and the trash profundity. On the off chance that the trash level is 70% or under 70%, at that point it's alright. Be that as it may, if the trash level is above 70% their Arduino gives data above container level to server ESP8266 01 module. A server is utilized to store information and shows all dustbin levels on the page.

GSM used to send an instant message to the versatile. The instant message contains data about the trash level and area of a specific container. This chapter on the IoT-based waste monitoring system is a very new system that helps to make the city clean. Through the help of the webpage, the capacity of squander gathered in the bin can be intimated easily to the administration. Ultrasonic sensors are situated inside the bins to locate the trash stage and balance it with the trash bin's intensity in this system. Arduino microcontroller, WiFi modem for sending data, and buzzers are used for the regulation of this system. The system is motorized by a 12 V transformer. The user can check the position of waste using a webpage. A picture view of the trash bins is given by the webpage and highlights the junk collected in colour to show the stage of trash collected.

5.7 PROPOSED SMART SOLUTION BASED ON MACHINE LEARNING

Our future proposed model will be founded on an application that empowers a ready when a cognizant resident out and about snaps an image of waste lying out and about or of an over-burden, tipping-over-the-street squander receptacle and sends it to the command centre utilizing a cross-stage good mobile application. The image got by the command centre

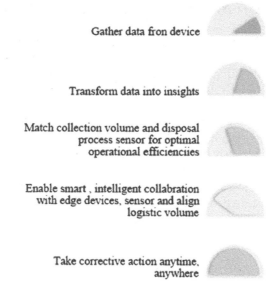

Gather data fron device

Transform data into insights

Match collection volume and disposal
process sensor for optimal
operational efficienciies

Enable smart , intelligent collabration
with edge devices, sensor and align
logistic volume

Take corrective action anytime,
anywhere

Figure 5.7 Machine learning model.

would then be able to be broken down utilizing a picture point and vector system analyzer to discover both the rough amounts just as the potential classes of the distinctive waste materials caught in the picture. The procedure stream requires no human mediation and it utilizes a smart algorithm to coordinate with past and existing information as shown in Figure 5.7.

The sensors joined to the side of the road squander canisters track squander assortment inside the receptacle and caution the waste assortment trucks naturally utilizing an IoT-coordinated and arranged framework, however, squander that is tossed straightforwardly out and about getaways the advanced visual eyes of the sensor cameras. It needs caution and a cognizant resident or human obstruction to cover that bit of the waste tossed out and about by an errant resident. The admonition here is that the caution and the cognizant resident must have fundamental information on photography and must be open to utilizing applications on cell phones. Information from both the sensors and the photos sent by an alarm and a cognizant resident is through a mind-boggling arrangement of multi-point and multi-layered examination. We will use past waste information to prepare the framework for a machine learning (ML) stage to recognize and arrange squander and roughly measure the heaviness of the waste. The ML stage will utilize past pictures of waste taken from more than 60 canister areas in the city, at different hours of the day. The ML stage is likewise prepared on standard things by and large found in and around squander receptacles so it can recognize them without any problem. Thus this project will be interesting for society [26–40].

5.8 CONCLUSION AND FUTURE SCOPE

IoT-based trash checking framework is an inventive framework that will keep the urban areas clean. This chapter gives a fundamental thought regarding a productive trash checking framework by utilizing the innovation of IoT. Through implementing this undertaking, we would abstain from flooding garbage from the compartment in a nearby location that is either physically piled beforehand or in customary trucks with the aid of loaders. It can naturally screen the trash level and send the data to the gathering truck. This works if the level of trash in the canister is more noteworthy than 80%, at that point, a warning will be sent to the vehicle to clean the trash container. The rate likewise can be seen by the experts with the assistance of a database and website page created. This can be reached out by making the framework worldwide and all the trash vehicles can know the status of the canister so that the closest vehicle with the assistance of dynamic directing calculation strategies can clean the receptacle quickly to improve the proficiency of the framework. This is extremely energizing the occasions we're living in, with AI and the AI infiltrating every aspect of our lives. I'm anticipating the improvement of this zone. For this, in the future, I will try to attach the camera with an MCU node so that this garbage system can also use for security purposes as well as the garbage monitoring system work as a Cyber-Physical System. Thus, ML-based systems will be a special gift to society.

REFERENCES

[1] Sathish Kumar, N., Vijaylakshmi, B., Jenifer Prathana, R., & Shankar, A. (2016). *IOT Based Smart Garbage alert system using Arduino UNO*. USA: IEEE.

[2] Thakker, S., & Narayanamoorthi, R. (2015). Smart and wireless waste management. In *IEEE Sponsored 2nd International Conference on Innovations in Information Embedded and Communication Systems*. USA.

[3] Borozdukhin, A., Dolinina, O., & Pechenkin, V. (2016). *Approach to the Garbage Collection in the Smart Clean City Project*. Saratov, Russia: Yuri Gagarin State Technical University.

[4] Anagnostopoulos, T., Zaslavs, A., Medvedev, A., & Khoruzhnicov, S. (2015). Top–k query based dynamic scheduling for IOT enabled smart city waste collection. In *16th IEEE International Conference on Mobile Data Management*. Singapore.

[5] Rana, A. K., & Sharma, S. (2021). The fusion of blockchain and IoT technologies with industry 4.0. In *Intelligent Communication and Automation Systems* (pp. 275–290). USA: CRC Press.

[6] Rana, A. K., Sharma, S., Dhawan, S., & Tayal, S. (2021). Towards secure deployment on the internet of robotic things: architecture, applications, and challenges. In *Multimodal Biometric Systems* (pp. 135–148). Singapore: CRC Press.

[7] Arora, S., Sharma, S., & Rana, A. K. (2021). 8 Ultrawide band antenna for wireless communications. *Internet of Things: Robotic and Drone Technology, 95.*

[8] Kumar, A., & Sharma, S. (2021). Demur and routing protocols with application in underwater wireless sensor networks for smart city. In *Energy-Efficient Underwater Wireless Communications and Networking* (pp. 262–278). USA: IGI Global.

[9] Rana, A. K., Sharma, S., Dhawan, S., & Tayal, S. (2021). 11 Towards secure. *Multimodal Biometric Systems: Security and Applications, 135.*

[10] Zaslavsky, A., & Georgakopoulos, D. (2015). Internet of things: Challenges and state-of-the-art solutions in internet-scale sensor information management and mobile analytics. In 16th IEEE International Conference on Mobile Data Management. USA.

[11] Vinothkumar, B., Sivaranjani, K., Sugunadevi, M., & Vijayakumar, V. (2015). IOT based garbage management system. *International Journal of Science and Research (IJSR), 6, 99–101.*

[12] Patil, S., Mohite, S., Patil, A., & Joshi, S. D. (2017). IoT based smart waste management system for smart city. *International Journal of Advanced Research in Computer Science and Software Engineering.* ISSN: 2277128X Vol. 7, Issue 4.

[13] Naregalkar, P. R., Thanvi, K. K., & Srivastava, R. (2017). IOT based smart garbage monitoring system. *International Journal of Advanced Research in Electrical, Electronics and Instrumentation Engineering.* ISSN (Online): 2278 8875 Vol. 6, Issue 5.

[14] Kulkarni, P., Patil, V., Chavan, A., Powar, R., & Dhaygude, V. (2017). GSM based garbage management system. *International Journal of Electrical and Electronics Engineers.* ISSN:2321-2055, Vol. 9, Issue 1.

[15] Rana, A. K., & Sharma, S. (2019). Enhanced energy-efficient heterogeneous routing protocols in WSNs for IoT application. *IJEAT, 9*(1), 4418–4415.

[16] Kumar, A., Sharma, S., Goyal, N., Singh, A., Cheng, X., & Singh, P. (2021). Secure and energy-efficient smart building architecture with emerging technology IoT. *Computer Communications, 176, 207–217.*

[17] Kumar, A., et al. (2021). Revolutionary strategies analysis and proposed system for future infrastructure in Internet of Things. *Sustainability, 14*(1), 71.

[18] Rana, A., Chakraborty, C., Sharma, S., Dhawan, S., Pani, S. K., & Ashraf, I. (2022). Internet of medical things-based secure and energy-efficient framework for health care. *Big Data, 10*(1), 18–33.

[19] Kumar, A., & Sharma, S. (2021). Internet of robotic things: design and develop the quality of service framework for the healthcare sector using CoAP. *IAES International Journal of Robotics and Automation, 10*(4), 289.

[20] Dhawan, S., Chakraborty, C., Frnda, J., Gupta, R., Rana, A. K., & Pani, S. K. (2021). SSII: secured and high-quality steganography using intelligent hybrid optimization algorithms for IoT. *IEEE Access, 9, 87563–87578.*

[21] Rana, S. K., Kim, H. C., Pani, S. K., Rana, S. K., Joo, M. I., Rana, A. K., & Aich, S. (2021). Blockchain-based model to improve the performance of the next-generation digital supply chain. *Sustainability, 13*(18), 10008.

[22] Rana, A. K., & Sharma, S. (2021). Internet of things based stable increased-throughput multi-hop protocol for link efficiency (IoT-SIMPLE) for health

monitoring using wireless body area networks. *International Journal of Sensors Wireless Communications and Control, 11*(7), 789–798.

[23] Kumar, A., Sharma, S., Goyal, N., Gupta, S. K., Kumari, S., & Kumar, S. (2022). Energy-efficient fog computing in Internet of things based on routing protocol for low-power and lossy network with Contiki. *International Journal of Communication Systems, 35*(4), e5049.

[24] Pandit, M. et al. (2022). Towards design and feasibility analysis of DePaaS: AI based global unified software defect prediction framework. *Applied Sciences, 12*(1), 493.

[25] Lilhore, U. K. et al. (2022). Enhanced convolutional neural network model for cassava leaf disease identification and classification. *Mathematics, 10*(4), 580.

[26] Rana, A. K., & Sharma, S. (2021). Industry 4.0 manufacturing based on IoT, cloud computing, and big data: manufacturing purpose scenario. In *Advances in Communication and Computational Technology* (pp. 1109–1119). Singapore: Springer.

[27] Rana, A. K., Krishna, R., Dhwan, S., Sharma, S., & Gupta, R. (2019, October). Review on artificial intelligence with Internet of Things-problems, challenges and opportunities. In *2019 2nd International Conference on Power Energy, Environment and Intelligent Control (PEEIC)* (pp. 383–387). USA: IEEE.

[28] Rana, A. K., & Sharma, S. (2021). Contiki Cooja security solution (CCSS) with IPv6 routing protocol for low-power and lossy networks (RPL) in Internet of Things applications. In *Mobile Radio Communications and 5G Networks* (pp. 251–259). Singapore: Springer.

[29] Palkar, P. M., Pathan, T., Hedaoo, A. P., Harode, K. A., Petkule, N. M., Kakade, P. P., & Kolhe, P. D. (2017). Smart city garbage collection monitoring system. *IJARIIE*. ISSN(O)-2395-4396, Vol. 3, Issue 2.

[30] Lokhande, P., & Pawar, M. D. (2017). Garbage collection management system. *International Journal of Engineering and Computer Science*. ISSN:2319-7242, Vol. 5, pp. 18800–18805.

[31] Arisz, H., & Burrell, B. C. (2006). *Urban Drainage Infrastructure Planning and Design Considering Climate Change*. USA: IEEE.

[32] Karadimas, D., Papalambrou, A., Gialelis, J., & Koubias, S. (2017). *An Integrated Node for Smart-City Applications Based on Active RFID Tags; Use Case on Waste-Bins*. IEEE. USA.

[33] Al-Masri, E., et al. Investigating messaging protocols for the Internet of Things (IoT). *IEEE Access*. 10.1109/ACCESS.2020.2993363

[34] Andrada Tivani, A. E., Murdocca, R. M., Sosa Paez, C. F., & Dondo Gazzano, J. D. (February 2020). Didactic prototype for teaching the MQTT protocol based on free hardware boards and node-RED. *IEEE Latin America Transactions, 18*(02), 376–382. 10.1109/TLA.2020.9085293

[35] Sofwan, A., Sumardi, S., Ahmada, A. I., Ibrahim, I., Budiraharjo, K., & Karno, K. (2020). Smart Greetthings: Smart Greenhouse Based on Internet of Things for Environmental Engineering. In 2020 International Conference on Smart Technology and Applications (ICoSTA), Surabaya, Indonesia, pp. 1–5. 10.1109/ICoSTA48221.2020.1570614124

[36] Jaikumar, K., Brindha, T., Deepalakshmi, T. K., & Gomathi, S. (2020). IOT assisted MQTT for segregation and monitoring of waste for smart

cities. In 2020 6th International Conference on Advanced Computing and Communication Systems (ICACCS), Coimbatore, India>, pp. 887–891. 10.1109/ICACCS48705.2020.9074399

[37] Kumar, A., & Sharma, S. (2021). IFTTT rely based a semantic web approach to simplifying trigger-action programming for end-user application with IoT applications. In *Semantic IoT: Theory and Applications* (pp. 385–397). Cham: Springer.

[38] Kumar, A., & Sharma, S. (2021). 11 Internet of Things. *Electrical and Electronic Devices, Circuits and Materials: Design and Applications, 183.*

[39] Kumar, A., Sharma, S., Dhawan, S., Goyal, N., & Lata, S. (2021). 14 E-learning with Internet of Things. *Internet of Things: Robotic and Drone Technology, 195.*

[40] Dhawan, S., Gupta, R., Rana, A., & Sharma, S. (2021). Various swarm optimization algorithms: review, challenges, and opportunities. *Soft Computing for Intelligent Systems, 291–301.*

Chapter 6

Application of Natural Language Processing and IoT to Emulate Virtual Receptionist

Amrita Singh and Rishabh Jha

CONTENTS

6.1 INTRODUCTION TO VIRTUAL ASSISTANTS

Virtual assistants are software that can perform tasks as per the immediate type of task or services required by the user. It is also known as intelligent personal assistant. Basically, a virtual assistant takes commands as a question and then performs the tasks as per the question. Nowadays a specific term is used for virtual assistants, called "Chat Bots" which are used for those hosted on the web. It is used for various purposes; from entertainment, education, marketing, sales representative, and as a substitute for service centre query response unit. Virtual assistants are able to interpret human voice signals and generate outputs accordingly. Using these assistants, users can control their devices, home appliances, and manage emails, calendars, and to-do-list using voice commands.

Artificial intelligence and the Internet of Things (IoT) are inextricably linked. Large volumes of data are generated by IoT systems, and data are at the heart of artificial intelligence and machine learning. Simultaneously, as the number of linked devices and sensors grows, so does the importance of smart technology in this industry [1–8]. Computer intelligence is now used in IoT goods in a variety of ways, depending on the needs. This chapter focuses on natural language processing (NLP), a subset of artificial intelligence. The ability to understand human speech is one of the most fundamental principles

DOI: 10.1201/9781003355960-6

in NLP. Voice control on many platforms is impossible to implement without NLP. It's difficult to overestimate the importance of speech recognition in the IoT [9–15]. The hands-free speech interface has a lot of potential in the IoT world. In some circumstances, it's simply a matter of usability; the more sophisticated the system, the more difficult it is to develop a user-friendly mobile or online interface to control it. As a result, the voice interface is simple to use and does not require extensive training [16–20].

As of now, the capabilities and smartness of virtual assistants have greatly increased. Enterprises such as Google, Microsoft, Amazon are using this technology with their devices. At the same time, even small enterprises are using these services using Chat Bots. Apple's Siri on IOS and Google's Assistant on Android are the game changers in the smartphone industry. Similarly, Microsoft's Cortana is popular on windows devices. Amazon's Alexa, Echo Dot, is also popular in speakers choice.

Virtual assistant's common usage is as follows:

- Chats in messengers either from person to person or person to enterprise.
- Virtual assistants are also used to take voice input commands in Google's Assistant on Android, Apple's Siri on IOS, Microsoft's Cortana, and Amazon's Alexa, Echo Dot.

Alphabet's Google uses virtual assistants in various forms, such as Google Assistant via chat on the Google Allo and Google Messages app and via voice on Google Home smart speakers.

6.2 PROBLEM STATEMENT

During the time of pandemic, the operational part of the organizations are almost operating from work-from-home mode but the front desk and reception area of the organizations are operating at the risk of viral infection. This makes the front desk area the super spreader of the virus. The tasks that these officers perform are:

- Sending emails for corporates
- Information sharing about the organization
- Making appointments
- Organization branding
- Contact number provider

If all these operations can be done by a robot powered by a smart virtual assistant then these tasks can be performed without risking anyone's life in this time of pandemic. The robot can perform all these tasks without any biases and in no time which can give third persons a great experience while visiting an organization for the first time.

6.3 OBJECTIVES

The basic objectives of the virtual receptionist are as follows:

- to digitalize the information rendering process of an organization,
- to deliver the information without any biases,
- to render information to the parties without any lag in time,
- to maintain the accuracy of the information provided to the people without having any manual errors, and
- to act as a brand influencer for an organization by providing information on what the owners want to deliver from the robot.

6.4 SCOPES AND LIMITATION

This project is willing to provide a next-gen information rendering system experience to the users. This wants to eliminate any manual errors due to the presence of biases in the human brain. On the other hand, standard virtual assistants have a lot of data and because of that the information rendered by them is not always as needed. That's one of the bases of the need for virtual assistants like CITY [21–25].

CITY, on the other hand, is trained with only data that concerns its organization, i.e., College of Information Technology and Engineering. The less the noise, the lesser will be error and the better will be the accuracy of the virtual receptionist – CITY. The product limitations of CITY are as follows:

1. The general perception of people about Robot is one that can move and have locomotion, but due to its primary objective in the 1st version, CITY does not have locomotive features.
2. It has no dedicated data storage channel in its 1st version to avoid any further complexity.

6.5 LITERATURE REVIEW

Virtual assistants have been playing a pivotal role in connecting virtual teams together by acting as the bridge between the team members who are distantly present. This helps to manage the remotely working teams to bring on the ideas on a common table. Similarly, virtual assistants have been working very well for bringing interests to students in teaching–learning process. The actual growth in the students in various domains in which they are engaged can be traced effectively by these assistants. Research has also shown that an even more proficient communication between various parties within and outside the companies can be achieved by these kinds of assistants. These assistants have been a boon for various applications in the field of health sector as well where they act as a mediator between patients and healthcare workers for clear and meaningful conversation [26–30].

6.6 FUNCTIONALITIES OF CITY

a. **Time check**

On the voice command about date and time, CITY gives both speech and text output of time and date as shown in Figure 6.1. It also greets according to time, good morning if it is morning and so on.

b. **Opening Google and YouTube**

Just like Google's Assistant and Amazon's Alexa, it can take user commands and open popular sites such as Google, Facebook, College websites, YouTube, and so on as shown in Figures 6.2 and 6.3, respectively.

c. **Scrapping Data from Wikipedia**

Using CITY, we can scrap data from Wikipedia. Whatever we want to search on Wikipedia we just have to ask CITY about that word including Wikipedia in the command, for example, "Sacred Games Wikipedia" will search Sacred Games in Wikipedia as shown in Figure 6.4.

d. **Sending corporate emails**

CITY also has the function of sending emails to the desired recipients. We have used smtplib, a pre-designed module of python for the accomplishment of this purpose of sending emails as shown in Figure 6.5.

e. **College Advertisement**

CITY can be used for personalized advertisements of firms. It can be explicitly programmed to provide the desired outputs to the customers. Moreover, human-to-human marketing is not as effective as robot-to-human conversation and humans believe robots even more than a human as shown in Figure 6.6.

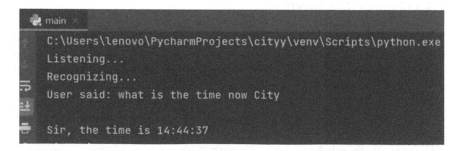

Figure 6.1 Displaying time.

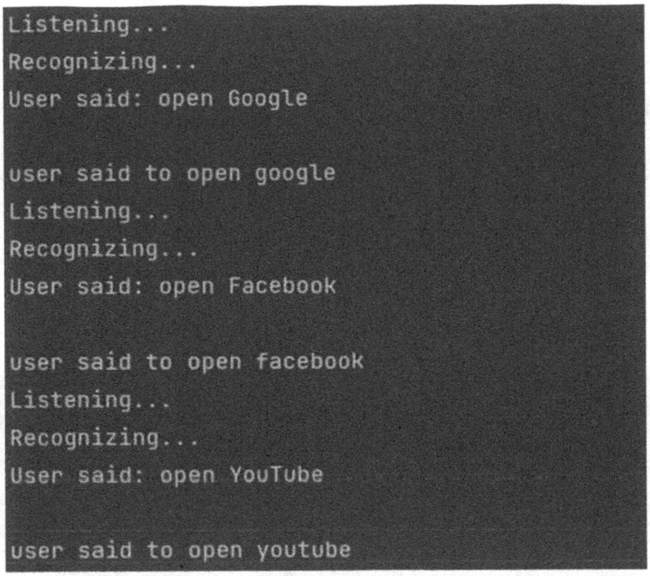
Listening...
Recognizing...
User said: open Google
user said to open google
Listening...
Recognizing...
User said: open Facebook
user said to open facebook
Listening...
Recognizing...
User said: open YouTube
user said to open youtube

Figure 6.2 Web browsing.

google.com
Google
Google Search
I'm Feeling Lucky

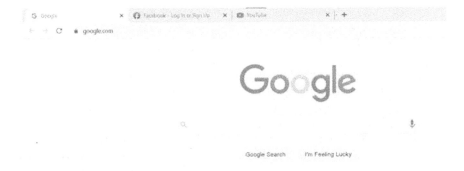

Figure 6.3 Displaying results.

Listening...
Recognizing...
User said: artificial intelligence Wikipedia
Artificial intelligence (AI) is intelligence demonstrated by machines, as opposed to natural intelligence displayed by animals including humans.
Leading AI textbooks define the field as the study of "intelligent agents": any system that perceives its environment and takes actions that maximi

Figure 6.4 User input for Wikipedia and output.

```
Listening...
Recognizing...
User said: send an email

Listening...
Recognizing...
User said: CIT College admissions are open till what date
```

Figure 6.5 User input for email.

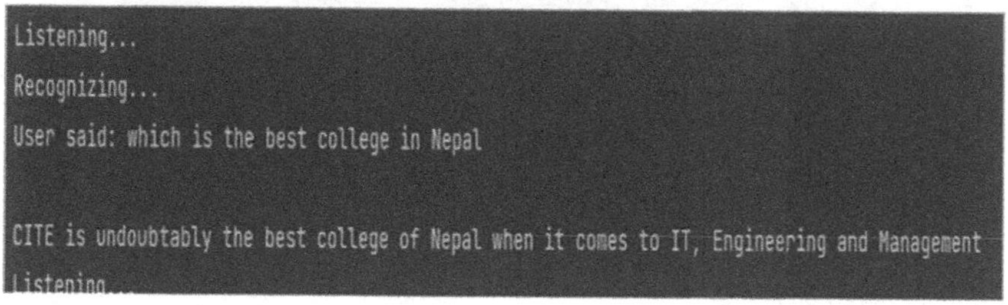

```
Listening...
Recognizing...
User said: which is the best college in Nepal


CITE is undoubtably the best college of Nepal when it comes to IT, Engineering and Management
Listening...
```

Figure 6.6 Displaying results.

6.7 CONCLUSION

The primary use of technology is to ease the life of people. City, the virtual receptionist uses the concept of NLP and deep learning to emulate the task performed by a receptionist in an organization. It is able to perform the tasks by eliminating the biases that are seen in tasks performed by manual receptionists. The accuracy in performing the tasks is greatly uplifted. With the growing use of AI in all domains of work, organized use of AI for specific purposes can be a game changer. Virtual receptionist, CITY is one such example where a specific problem is tried to solve using AI. This project has aimed to minimize the manual burden as well as tried to increase the efficiency and digitalize the process of information rendering in any organization. This project also aims to reduce the biases due to the involvement of manual sentiments.

REFERENCES

[1] Rafailidis, D., & Manolopoulos, Y. (2019, June). Can virtual assistants produce recommendations? In *Proceedings of the 9th International Conference on Web Intelligence, Mining and Semantics* (pp. 1–6). USA.

[2] Palkar, P. M., Pathan, T., Hedaoo, A. P., Harode, K. A., Petkule, N. M., Kakade, P. P., & Kolhe, P. D. (2017). Smart city garbage collection monitoring system. *IJARIIE*. ISSN(O)-2395-4396, Vol. 3, Issue 2.

[3] Lokhande, P., & Pawar, M. D. (2017). Garbage collection management system. *International Journal of Engineering and Computer Science*. ISSN:2319-7242, Vol. 5, pp. 18800–18805.

[4] Arisz, H., & Burrell, B. C. (2006). *Urban Drainage Infrastructure Planning and Design Considering Climate Change*. USA: IEEE.

[5] Karadimas, D., Papalambrou, A., Gialelis, J., & Koubias, S. (2017). An integrated node for Smart-City applications based on active RFID tags; Use case on waste-bins. IEEE. Singapore.

[6] White, R. W. (2018). Skill discovery in virtual assistants. *Communications of the ACM*, *61*(11), 106–113.

[7] Al-Masri, E., et al. Investigating messaging protocols for the Internet of Things (IoT). IEEE Access. 10.1109/ACCESS.2020.2993363

[8] Andrada Tivani, A. E., Murdocca, R. M., Sosa Paez, C. F., & Dondo Gazzano, J. D. (February 2020). Didactic prototype for teaching the MQTT protocol based on free hardware boards and node-RED. *IEEE Latin America Transactions*, *18*(02), 376–382. 10.1109/TLA.2020.9085293

[9] Sofwan, A., Sumardi, S., Ahmada, A. I., Ibrahim, I., Budiraharjo, K., & Karno, K. (2020). Smart Greetthings: smart greenhouse based on internet of things for environmental engineering. 2020 International Conference on Smart Technology and Applications (ICoSTA), Surabaya, Indonesia, pp. 1–5. 10.1109/ICoSTA48221.2020.1570614124

[10] Jaikumar, K., Brindha, T., Deepalakshmi, T. K., & Gomathi, S. (2020). IOT assisted MQTT for segregation and monitoring of waste for smart cities. 2020 6th International Conference on Advanced Computing and Communication Systems (ICACCS), Coimbatore, India, pp. 887–891. 10.1109/ICACCS48705.2020.9074399

[11] Al-Masri, E., et al. Investigating messaging protocols for the Internet of Things (IoT). IEEE Access. 10.1109/ACCESS.2020.2993363.

[12] Kumar, A., & Sharma, S. (2021). IFTTT rely based a semantic web approach to simplifying trigger-action programming for end-user application with IoT applications. In *Semantic IoT: Theory and Applications* (pp. 385–397). Cham: Springer.

[13] Kumar, A., & Sharma, S. (2021). 11 Internet of Things. *Electrical and Electronic Devices, Circuits and Materials: Design and Applications*, *183*.

[14] Kumar, A., Sharma, S., Dhawan, S., Goyal, N., & Lata, S. (2021). 14 E-learning with Internet of Things. *Internet of Things: Robotic and Drone Technology*, *195*.

[15] Dhawan, S., Gupta, R., Rana, A., & Sharma, S. (2021). Various swarm optimization algorithms: Review, challenges, and opportunities. *Soft Computing for Intelligent Systems*, 291–301.

[16] Rana, A. K., & Sharma, S. (2021). Internet of Things based stable increased-throughput multi-hop protocol for link efficiency (IoT-SIMPLE)

for health monitoring using wireless body area networks. *International Journal of Sensors Wireless Communications and Control, 11*(7), 789–798.

[17] Kumar, A., Sharma, S., Goyal, N., Gupta, S. K., Kumari, S., & Kumar, S. (2022). Energy-efficient fog computing in Internet of Things based on Routing Protocol for Low-Power and Lossy Network with Contiki. *International Journal of Communication Systems, 35*(4), e5049.

[18] Pandit, M., Gupta, D., Anand, D., Goyal, N., Aljahdali, H. M., Mansilla, A. O., ... & Kumar, A. (2022). Towards design and feasibility analysis of DePaaS: AI based global unified software defect prediction framework. *Applied Sciences, 12*(1), 493.

[19] Lilhore, U. K., Imoize, A. L., Lee, C. C., Simaiya, S., Pani, S. K., Goyal, N., ... & Li, C. T. (2022). Enhanced convolutional neural network model for Cassava leaf disease identification and classification. *Mathematics, 10*(4), 580.

[20] Rana, A. K., Sharma, S., Dhawan, S., & Tayal, S. (2021). 11 Towards secure. *Multimodal Biometric Systems: Security and Applications, 135.*

[21] Zaslavsky, A., Georgakopoulos, D. (2015). Internet of Things: Challenges and state-of-the-art solutions in Internet-scale sensor information management and mobile analytics. 16th IEEE International Conference on Mobile Data Management. USA.

[22] Vinothkumar, B., Sivaranjani, K., Sugunadevi, M., & Vijayakumar, V. (2015). IOT based garbage management system. *International Journal of Science and Research (IJSR), 6*, 99–101.

[23] Patil, S., Mohite, S., Patil, A., & Joshi, S. D. (2017). IoT based smart waste management system for smart city. *International Journal of Advanced Research in Computer Science and Software Engineering.* ISSN: 2277128X, Vol. 7, Issue 4.

[24] Naregalkar, P. R., Thanvi, K. K., & Srivastava, R. (2017). IOT based smart garbage monitoring system. *International Journal of Advanced Research in Electrical, Electronics and Instrumentation Engineering.* ISSN (Online): 2278 8875, Vol. 6, Issue 5.

[25] Kulkarni, P., Patil, V., Chavan, A., Powar, R., & Dhaygude, V. (2017). GSM based garbage management system. *International Journal of Electrical and Electronics Engineers.* ISSN: 2321-2055, Vol. 9, Issue 1.

[26] David, B., Chalon, R., Zhang, B., & Yin, C. (2019, May). Design of a collaborative learning environment integrating emotions and virtual assistants (chatbots). In *2019 IEEE 23rd International Conference on Computer Supported Cooperative Work in Design (CSCWD)* (pp. 51–56). IEEE.

[27] Reyes, R., Garza, D., Garrido, L., Cueva, V. D. L., & Ramirez, J. (2019, October). Methodology for the implementation of virtual assistants for education using Google dialogflow. In *Mexican International Conference on Artificial Intelligence* (pp. 440–451). Cham: Springer.

[28] Rana, A. K., & Sharma, S. (2021). ContikiCooja Security Solution (CCSS) with IPv6 routing protocol for low-power and lossy networks (RPL) in Internet of Things applications. In *Mobile Radio Communications and 5G Networks* (pp. 251–259). Singapore: Springer.

[29] Gubareva, R., & Lopes, R. P. (2020). Virtual assistants for learning: A systematic literature review. *CSEDU*, *1*, 97–103.

[30] Graham, C., & Jones, N. B. (2016). Intelligent virtual assistant's impact on technical proficiency within virtual teams. *International Journal of Virtual and Personal Learning Environments (IJVPLE)*, 6(1), 41–61.

Chapter 7

An Overview: Understanding of Image Fusion

Navneet Kaur and Nidhi Chahal

CONTENTS

7.1 INTRODUCTION

As more and more activities such as medical imaging, security, avionics, surveillance, and night vision use multimodal imaging arrays, multisensory image fusion has become an active topic of research. Similar arrays provide a wider spectral content and more reliable information in poor environmental conditions, but at the cost of a significant increase in data volume. Picture emulsion reduces data burden by fusing visual information from several image signals into a single-fused image, with the clear goal of preserving the multisensory information's full content value.

There has been a slew of image emulsion algorithms proposed [1–7] that layout various multimodal information preservation bundles. Furthermore, due to the significant data reduction involved in the emulsion process, some

DOI: 10.1201/9781003355960-7

algorithms commonly inject specific remnants, or erroneous information, into the fused image [2,4,5]. A reliable system for selecting the best emulsion algorithm for each operation is still mostly a work in progress. So far, only a few hangouts and emulsion evaluation criteria [8–11] are interesting because they have no previous ground verity (akin to an immaculately fused image, obtainable only in unusual instances [3,4]) have been offered. The grade representation metric [8] is based on the idea of determining the degree to which input grade information is preserved locally in the fused image. Another option is to leverage collective information between the inputs and the fused image, which has been proven to be reliable when dealing with multisensory data [9]. A more targeted technique based on assessing the similarity between the inputs and the fused image through original statistics was presented [10], as such a global approach is still sensitive to noise. Finally, in [11,12], the idea of visible differences was proposed for evaluating emulsion performance as the total area of apparent differences between the inputs and the fused image.

All of these criteria provide only a rudimentary numerical approximation of emulsion performance in terms of global similarity between the inputs and the fused image [9–11]. As a result, their scores only provide a hazy picture of the relative merits of various emulsion schemes in each operation. In general, the emulsion is a complicated process of information transfer and representation, and certain components of this process are more crucial in specific processes and may require special attention. For example, a loss of a critical moment during the emulsion of a medical image may result in a faulty opinion and more significant effects. An emulsion artefact created into the fused image by the emulsion process could cause a benign item to be identified as problematic or a valid target in military avionics. A considerably more extensive emulsion evaluation is required to address similar operation-specific features of emulsion performance robustly. This work presents a system for determining comprehensive, objective image emulsion performance based on an assessment of major information components such as emulsion gain, emulsion information loss, and emulsion remnants in image emulsion information benefits by each detector (artificial information created during emulsion).

7.2 EVOLUTION OF IMAGE FUSION

During the 1980s, the introduction of multisensory applications, particularly in the field of remote detecting, combined with broad examination revelations in pyramid-based change strategies, revealed picture combination as an exploration region for the securing of larger pictures for human representation [13].

As an examination theme, picture combination can be extraordinarily seen in two ways: first and foremost deliberation insightful, as a sub-branch of

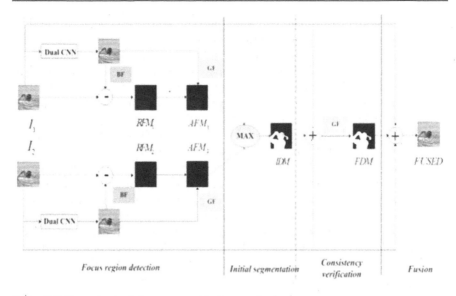

Figure 7.1 Flow chart of the proposed fusion method.

information combination, which additionally incorporates a combination of different information types such as sound, video, multifaceted, and complex mathematical information [14]. A few models have been incorporated, in particular an interruption discovery framework on the Internet that wires network information. Area gauges from a global positioning system (GPS) chip are fused with the vehicle's installed diagnostics (OBD) arrangement for the automated routes in modern automobiles [14]. Electroencephalography (EEG) results are combined with electrooculography (EOG) and respiratory indicators in the treatment of individuals with weakness [15].

Calculation savvy, the combination is an augmentation of picture examination techniques that additionally contain other picture handling instruments such as pressure and coding, including extraction, enrolment, acknowledgement, and division. All things considered, large numbers of the picture combination approaches examined here can for the most part be applied to other picture and sign handling applications. Models incorporate ICA for blind source division of EEG and ECG signals in medication [16] and facial acknowledgement [17] and the wavelet change for dimensionality decrease for use in picture coding.

A. Early Fusion Systems:

Originally, the combination's primary function was limited to human vision and autonomous direction. Pixel averaging is the quickest and most important type of combining, in which each pixel of all information pictures is individually summed and their usual pixel esteem is combined into the merged picture. Regardless, this technique is extremely rudimentary, and the outcomes have been unsuitably proven. The averaging technique presents ancient rarities particularly when elements present in just one info picture are "superimposed" on the intertwined yield, as can happen in visual numerous openings. It likewise causes design crossing out and differentiates decrease for the situation where two information sources have highlights of equivalent striking nature yet inverse difference.

Most methodologies are in this manner classified under middle-level combination (ILF), likewise called a combination of highlights as the interaction includes removing significant elements from the picture utilizing procedures, for example, multigoal investigation (MRA) and sign disintegration [18]. Organizes picture combination calculations into spatial and change area techniques in [19]. Since then, practically all combination computations have been based on a certain type of change region, in which each information image is changed and the change coefficients are combined. The reverse modification of the coefficients is used to create the composite image that follows. The main idea behind change-based combination algorithms is to modify the size of the source picture coefficients to amplify edges and angles.

B. *Pyramid-based Methods:*

Pyramid disintegration is a more substantial method of dealing with picture combining. A picture pyramid is a sort of multiresolution investigation (MRA) that consists of a collection of separated and scaled representations of a scene. The combination is done by determining coefficients at each scale from the source picture pyramids, followed by the opposite alteration of the next pyramid [20].

The low-pass Laplacian pyramid for the binocular combination was initially proposed in 1984) [21], who presented the pyramid approach. A Laplacian pyramid is similar to a Gaussian pyramid in that it is created by subtracting two progressive low-pass Gaussian pyramid levels. The level is scaled to a percentage of two from its previous level in a proportion of low-pass (RoLP) pyramid, whereas contrast pyramids are similar to RoLP but measure the proportion of glow of a specific place within a picture to the neighbourhood foundation luminance. Finally, a significant number of pyramid-based schemes have been presented and used in combination writing, including channel, deduct destroy (FSD), morphological, and angle pyramids [22].

In 1988, the main application of a combination of visible, warm, and infrared images [23] suggested that neural organizations be used to combine these modalities. The tremendous overhead necessary in handling complete photos is a flaw in the neural organizing technique. MRA techniques overcome this by degrading images into subtleties and normal channels, which can then be combined in the wavelet coefficient space. Keeping this in mind, pyramid disintegration techniques have become unavoidable in the picture combining field, although they are not without their drawbacks. It was not surprising to find blocking curios in interlaced photos, especially in locations where the multi-tangible information is essentially divergent in manner. Another problem with pyramids is their lack of adaptability, such as the lack of anisotropy and directional data.

C. *Wavelet-Based Methods:*

In 1993 a wavelet-based picture combination approach [24] acquainted us with. From there on in 1995 a review [25] likewise involved wavelets as elective premise capacities for multisensory picture combination, which can conquer the impediments of pyramid-based plans under its directionality. This implies a picture is decayed into its low-recurrence guess and flat, vertical, and slanting edges. Wavelets empower spatial data to be consolidated into the change interaction. These endeavours and another work [26] launched a pattern of utilizing wavelet changes for picture combination, on which a larger part of current combination calculations in presence is based.

Instead of the non-nearby and non-limited sinusoidal portrayals utilized by Fourier examination, wavelets can be appropriately utilized in endless areas and are ideal for approximating information with sharp discontinuities or edges. The wavelet change deals with a similar reason as the pyramid, where the sub-band coefficients of the comparing recurrence content are combined and the ensuing backward change creates an incorporated melded picture. Wavelet strategies that depend on fundamentally examining picture signals, in particular the discrete wavelet change (DWT), experience the ill effects of shift fluctuation by which an intermittence by a source sign could unfavourably influence its change same. This is an immediate outcome of the down-inspecting activity of wavelet changes. An option in contrast to this is to utilize oversampled plans, however, that would build excess and add to the handling time and cost. A more doable arrangement was proposed [27] in 1997 for a shift-invariant DWT (SI-DWT) technique which disposes of the subsampling step, subsequently delivering it overcomplete. Further, [28] presented a repetitive wavelet change (RWT), utilizing an undecimated type of the dyadic channel tree which is best executed through calculation. Somewhere else the wavelet addresses the shift variation DWT by bypassing the down-examining move in the disintegration cycle and using a bunch of new channels all through the course of every decay.

The improvement of complex-based wavelet changes, in particular, the doubletree complex wavelet change (DT-CWT) had the option to beat poor directional and recurrence selectivity issues encompassing past wavelet models, as well as diminishing over-fulfilment and effectively accomplishing amazing remaking. Its intricate property implies the stage data got from the change can be used for additional investigation if vital. Branch-offs got from the wavelet change incorporate contourlet, edge let, and curvelet change that joins anisotropic conduct and directional aversion to more readily work with the examination of fundamental picture highlights like edges.

D. Data-Driven Methods:

An apparent shortcoming of the wavelet change, and comparative changes such as Fourier and Gabor, is the consistent reliance of the premise capacities on a decent numerical property that bears no connection genuinely to the info information nearby, which regularly are non-straight and non-fixed. In such a manner, free part investigation (ICA), observational mode deterioration (EMD), and other non-parametric and information-driven techniques are viewed as prevalent as their highlights are straightforwardly gotten from the preparation of information. For example, rather than a standard bases framework utilizing wavelets, a bunch of bases that are appropriate for specific sorts of pictures might be prepared for ICA.

Conversely, EMD is an altogether versatile combination approach that makes no suppositions of the information. EMD works in the spatial space where it recursively deconstructs a picture into natural mode capacities (IMF) at various frequencies. The disintegration strategy uses envelopes related to the neighbourhood maxima and minima individually. The combination happens through a weighted blend of IMFs of information pictures. Contrasted with wavelet coefficients, the IMFs are not fixed and can be fit to fit the current information.

ICA and EMD are the two instances of non-parametric relapse that require a bigger example size of information, which thus supplies the model construction and model assessments. The deals with information displaying have since provoked a review into utilizing organically propelled models, to be specific shading vision for combination in 1995 [29], in which adversary handling was applied on intertwining noticeable and infrared pictures. From that point forward, new bearings in picture combination have prompted the improvement of a few organic models, which depend on how the human mind cycles and consolidates data got by various faculties. ICA, for instance, expects the answer for the "mixed drink party" issue in which hearable sign sources are recognized by the human mental framework.

E. State-of-the-Art in Image Fusion:

One more arising pattern in picture combination is the use of district-based techniques. This methodology is borne from the arrangement that districts, or healthy items inside a picture, will quite often convey data of interest. It, in this manner, seems OK to zero in on areas instead of simply individual pixels, since pixels can be handled all the more effectively assuming they are treated as an aggregate gathering inside a district as opposed to isolating elements. The district-based combination may accordingly assist with defeating a few disadvantages of pixel-based combination, such as obscuring, powerlessness to the commotion, and misregistration.

Other remarkable methodologies for picture combination incorporate those dependent on measurable and assessment hypotheses, as first proposed [30] utilizing Bayesian combination. Molecule models and Bayesian-based combinations accomplish predominant execution with high necessity applications; however, this frequently includes some significant pitfalls of higher computational intricacy. There are many different picture combination tactics available today, each with its own set of advantages and applicability for various purposes. Huge improvements have been made to bridge the gap between PC vision and human perceptions of image quality. Regardless, the pursuit of the best level of implantation quality continues (Figure 7.2).

7.2.1 Image Fusion Applications

It is by and large recognized that all imaging applications that involve the investigation of various picture data sources might profit from picture combinations. To be sure, picture combination is extremely valuable in an assortment of basic applications such as remote detecting, clinical imaging, modern deformity discovery, and military reconnaissance.

Remote sensing (RS) applications are concerned with the protection of geospatial images captured by satellites and airborne sensors such as SPOT, Quick Bird, IKONOS, and IRS. For applications such as urban

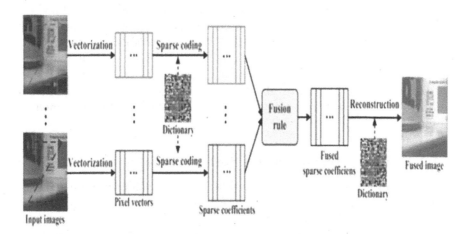

Figure 7.2 The main stages for a generic sparsity-based image fusion method.

planning, agribusiness, and topography, RS entails delivering top-notch geographic images in terms of both spatial and conceptual goals. Because of factors such as the sensor's radiation energy consumption and the limited information transfer rate from the satellite stage to the ground, developing an elite presentation sensor camera to execute such tasks is impossible. Rather, signal processing techniques are employed to achieve fairly excellent results.

A point-by-point survey of RS combination tactics was introduced in [31]. A panchromatic (PAN) picture depicting the scene in a high spatial goal but a single recurrence, and a multispectral/hyperspectral (MS/HS) picture capturing the scene in a huge number of ghostly goals across the frequency spectrum but at 1:4 the spatial goals of PAN. By integrating the nitty-gritty spatial goals of PAN into a resampled reproduction of multispectral photos using approaches like the wavelet change, the combination gives a down-to-earth and practical strategy to support detecting things with outstanding data from RS symbolism.

The power tint immersion (IHS) technique, in which the red-green-blue (RGB) hued space of the first MS symbolism is changed into IHS to get a better division of shading for combination with PAN pictures, is also an old-style combination strategy in RS applications, but it frequently results in ghastly corruption.

In medicine, picture fusion and other new advancements are increasingly being relied on for patient diagnosis and treatment [32]. A summary of clinical picture combinations was provided. The combination aids clinical imaging by providing a free composite of various image designs originating from various modalities, such as ultrasound, magnetic resonance imaging (MRI), computed tomography (CT), positron emanation tomography (PET), and single-photon discharge processed tomography (SPECT), which helps with outlining and separating designated objects of interest such as growths and veins. A therapy plan for radiotherapy in radiation oncology incorporates CT information mostly for patients' part computation, but cancer blueprints are best addressed in MRI. CT best reveals denser tissues with low contortion for clinical diagnosis, MRI provides more comprehensive data on fragile tissues with higher twisting, and PET provides better data on the bloodstream with a generally low spatial objective [9]. Using picture combinations can help you recognize important physical objects of interest from both sources.

7.2.2 Gradient-Based Fusion Performance

The inclination-based, objective combination assessment presented in [8] is used for the combination depictions proposed in this paper. By evaluating the accomplishment of angle data move from contributions to the interlaced image, the QAB/F system [8] couples substantial visual data with slope data and surveys. Combination calculations that accurately shift

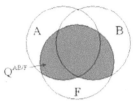

Figure 7.3 Graphical representation of the image.

more information inclination data into the entangled image are believed to perform better.

7.2.3 Fusion Performance Characterizations

The course of data combination should be evident as a data move problem in which at least two datasets are merged into another that should have all of the data from the initial sets in its default setting. In Figure 7.3, this is depicted graphically using a simple Venn chart. On account of picture signal combination data is connected to a spatial area that can each be viewed as a singular dataset. During combination, input pictures A and B are consolidated into another intertwined picture F by moving, preferably all of their data into F. The data effectively moved, displayed in Figure 7.3 as the concealed crossing point among A and B on one side and F, is exactly what the angle combination assessment system $Q^{AB/F}$ measures, comparative with the all-out size of A‰B. By and by, in any case, not all of the accessible information data is fundamentally moved into the melded picture and a few losses of data from the two sources of info happen. Simultaneously, the combination interaction itself incidentally makes extra, bogus data, known as combination antiquities in the intertwined picture. To present a comprehensive picture of the data blending process, it is necessary to assess these effects as well as provide a breakdown of the commitments in the effectively merged data provided by each sensor as well as data that is normal and selected for both of them.

7.2.4 Individual Sensor Contribution

Commitments in data towards the combined picture made by every sensor (input picture) in the combined cycle are expressly assessed in the $Q^{AB/F}$ system portrayed in Section 7.2. Angle data safeguarding gauges Q^{AF} and Q^{BF} are given by condition [3], which addresses how much data moved from A and B separately into F at every area (n, m). Complete data commitments of each contribution towards F are then assessed by condition [5] (for one or the other A or B) as the amount of all nearby commitments adjusted by the proportion of neighbourhood perceptual significance (WA and WB [8]). The commitments of individual contributions to the combined interaction are likewise represented in Figure 7.4.

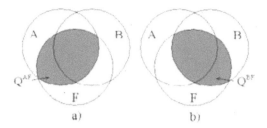

Figure 7.4 Information contribution.

7.2.5 Fusion Gain: Exclusive Information

Individual contributions to the combined cycle commitments Q^{AF} and Q^{BF} include both data that is specific to each piece of information and data that is common to all of them. The prior class is particularly important since it covers combination gain, or the benefit of combining data, which would otherwise be unavailable to an observer with only individual sources of information. This sum is depicted in Figure 7.5a and consists of two separate commitments of restricted data from each of the data sources.

The assessment of combination gain from the embraced structure isn't clear and requires a full breakdown of the commitments made by the information pictures. At first, neighbourhood restrictive data in F, Qå is found as the outright distinction of the slope data conservation coefficients Q^{AF} and Q^{BF}, condition [6]. This evaluates the aggregate sum of nearby restrictive data across the melded picture. For areas with a solid connection between the data sources, Qå will be little or zero, demonstrating no restrictive data. Then again, in regions where one of the sources of info gives a significant element that is absent in the other, this amount will tend towards 1.

7.2.6 Fusion Loss

Combination misfortune $L^{AB/F}$ is a proportion of the data lost during the combination interaction. This is the data accessible in the info pictures

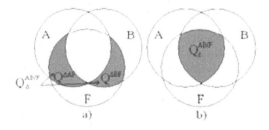

Figure 7.5 Exclusive information fused from both.

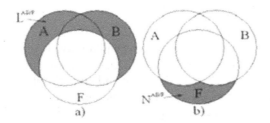

Figure 7.6 Fusion information loss.

however not in the melded picture. Combination misfortune is delineated graphically in Figure 7.6a.

An immediate loss of data is recognized as Q^{AF} and Q^{BF} esteems lower than 1, in any case, to assess combination misfortune appropriately one should have the option to recognize it from combination antiques, see 3.5, which likewise bring about Q^{AF} and $Q^{BF} < 1$. The $Q^{AB/F}$ system separates involving relative angle qualities in the sources of info and intertwined pictures, condition [1]. Henceforth, the accompanying arrangement is made for every area: assuming inclination strength in F is bigger than that in the data sources, F contains ancient rarities; then again, a more fragile angle in F demonstrates a deficiency of information data. The absolute combination data misfortune is in this manner assessed as a perceptually weighted nearby combination misfortune, given as $1 - Q^{AF}$ and $1 - Q^{BF}$ for the two information sources A and B, condition [33], incorporated over where the sign slope is more grounded in the information sources contrasted with the melded picture, for example where rn,m banner is 1, condition [34].

7.2.7 Fusion Artefacts

Combination of ancient rarities addresses visual data brought into the intertwined picture by the combination cycle that has no related high-lights in any of the sources of info. Combination curios are bogus data that straightforwardly cheapens the helpfulness of the melded picture and can have genuine outcomes in specific combination applications. Combination antiques are distinguished graphically as the data restrictive to F in Figure 7.6b. Inside the embraced system combination antiquities can be assessed as angle data that exists in F yet not in any of the info pictures. By and by, nearby gauges of combination antiques or combi-nation clamour as they are here and there called, Nn,m are assessed as combination misfortune where melded inclinations are more grounded than input, condition [18]. All out-combination antiques for the combi-nation interaction A, B o F are assessed as perceptually weighted in-corporation of the combination commotion gauges over the whole intertwined picture, condition [13].

7.3 METHODOLOGY

A broad outline of the field of picture combination is introduced. The concentrate first and foremost digs into the issue of numerous modalities that structure the inspiration for combination and talks about the primary benefits of picture combination. Further, it examines exhaustively the historical backdrop of combination calculations that contain different change areas and information-driven techniques. A part on picture combination applications, going from geospatial, clinical to security fields, is additionally introduced. By and large, this research straightforwardly plans to expose the advances and best in class inside the picture combination research region to help different fields. A helpful strategy in different utilizations of remote detecting includes the combination of panchromatic and multispectral satellite pictures. As of late, the presented strategy is a quick power tone immersion (IHS) combination technique. Besides its quick registering ability for intertwining pictures, this strategy can stretch out customary three-request changes to a subjective request. It can likewise rapidly blend gigantic volumes of information by requiring just resample multispectral information. Notwithstanding, quick IHS combination additionally misshapes shading similarly as combination cycles like the IHS combination procedure. This Image Fusion approach empowers quick, simple execution. Moreover, the tradeoff between the spatial and otherworldly goals of the picture to be intertwined can be effortlessly controlled with the guide of the tradeoff boundary.

A. *Fusion for Multimodality Sensors:*

In true applications where different optical sensors are utilized for picture procurement, it is regularly hard to acquire a decent-quality picture from a solitary sensor alone. Choices about framework conditions are seldom made upon the estimation of a solitary boundary. This condition stays valid across many parts of present-day innovation be it medication, topography, or defense.

The word 'excellent quality will encompass a variety of aspects of the image landscape, including brightness, sharpness, agitation, and distinctiveness, among others. There are numerous sensor technologies available, including the optical camera, millimeter-wave (MMW) camera, infrared (IR) and near-infrared (NIR), X-ray, radar, and magnetic resonance imaging (MRI), among others, each of which will highlight a different section of a captured image. Aside from sensor modalities, the pluralistic concept of information pictures is also important due to a variety of factors, including obstruction of objects of interest due to smoke, haze, and other undesirable particles, changing brightening in the landscape for photography applications, such as sunlight openness at different times of day, and movable boundaries within the actual sensors, such as the central length (Figure 7.7).

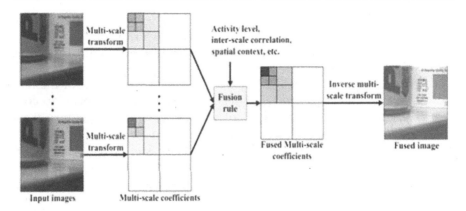

Figure 7.7 The main stages for a generic multiscale-based image fusion method.

- Regardless of the abovementioned, a lot of information will generally add to issues normal in signal handling including capacity necessities, calculation time, restricted transfer speed, and irregularity with choices about sensor frameworks [35], just as the absence of standard evaluation rules to gauge sensors having a place with various modalities. It, consequently, seems OK to decrease this complex information into simply a minimal single picture that jelly pertinent data and whose quality surpasses any of its bits of feedback. The fruitful converging of differentiating, yet integral, highlights from numerous sensors ought to in this manner be the objective of the picture securing framework.

- The UN Camp setup depicted in Figure 7.3 is a simple model. Two image acquisition methodologies are used to capture a scene in the evening. A NIR camera that detects a solid warm presence (in the 1.5–15 m region, for example) contains one piece of information. NIR sensors, on the other hand, suffer from reduced picture objective, main picture commotion, and lack of access to informational collections, all of which make them unsuitable for single use [33]. The following data is a standard photograph of a similar scene obtained with a visual camera that captures solid textural foundation subtleties (in the 0.45–0.7 m range) but is severely limited in low-light settings. The goal in this situation is to improve the lighting conditions in the environment and further enhance picture qualities, as well as work with the discovery of various moving objects and the detachment of pre-indicated objects of interest (tracking).

- Contingent upon client necessities, a decent picture should have the capacity to recognize the human figure against the definite scenery of landscape and ranger service. The arrangement of a decent-quality picture is vital as it empowers us to have a legitimate comprehension of the landscape setting, which might demonstrate unequivocal in true reconnaissance and target acknowledgement frameworks.

- In that lays the idea of combination. As a component of the fantastic test in picture handling, picture combination plans to consolidate the notable parts of at least two source pictures from these sensors to create a solitary result picture, that contains all appropriate picture highlights [34] and has a higher clear line of sight and mathematical quality than any info, which might be fundamental in basic applications like military observation. For this situation, a basic relationship of pixels won't get the job done because of the very different modalities utilized for picture procurement.
- The motivation behind the combination rises above the result picture quality alone. It empowers clients to picture various arrangements of the information under one scene. A significant variable why combination has been so effective is that architects, engineers, and clients can save costs by using signal handling procedures of planning a costly framework for picture obtaining. Combination decreases information dimensionality while protecting striking data content, in this manner lessening stockpiling costs [18] (Figure 7.8).

Tangible frameworks [33] centre principally around removing data from tactile information. The combination is in this manner wanted to work on

Figure 7.8 Example of visual-NIR image fusion.

visual exactness and suggest explicit inductions that couldn't be accomplished by a solitary sensor. The combination structure takes after how people find their environmental elements by utilizing different prompts from numerous modalities. For example, people use binocular vision by which they consolidate visual substance from both the left and right eyes for visual handling.

The combination of image fusion techniques likewise further develops framework dependability by diminishing vulnerability in factors, subsequently expanding precision [18], and supporting situational mindfulness [33]. It empowers better direction, confinement, and segregation of objects of interest – because of the probability of more complete data [33]. Above all, it improves and empowers us to recognize corresponding data for identification and division in numerous security applications.

B. Existing Image Fusion Techniques:

During the 1980s, picture combination got huge consideration from specialists in remote detecting and picture handling, as SPOT 1 (sent off in 1986) gave high-goal (10 m) Pan pictures and low-goal (20 m) MS pictures. Since that time, much exploration has been done to foster powerful picture combination procedures. For instance, an audit paper on picture combination strategies, distributed in 1998 in the *International Journal of Remote Sensing*, referred to roughly 150 scholastic papers on picture combination. Beginning around 1998, further logical papers on picture combination have been distributed with the accentuation on further developing combination quality and diminishing shading contortion. Among the many varieties of picture combination strategies, the most well-known and powerful are, for instance, IHS (Intensity, Hue, Saturation), PCA (principal components analysis), math blends, and wavelet base combination.

- The IHS combination converts a shading picture from RGB (Red, Green, and Blue) to IHS shading space. Because the force (I) band resembles a Pan picture, it is replaced in the combo with a high-goal Pan picture. The Pan and the tint (H) and immersion (S) groups are then subjected to an opposing IHS modification, resulting in an IHS combined image.

- The PCA change changes over intercorrelated MS groups into another arrangement of uncorrelated parts. The primary part additionally takes after a Pan picture. It is, in this way, supplanted by a high-goal Pan for the combination. The Pan picture is intertwined into the low-goal MS groups by playing out an opposite PCA change.

- Diverse math blends have been produced for picture combinations. The Transforms, SVR (Synthetic Variable Ratio), and RE (Ratio Enhancement) procedures are a few effective models. The essential technique of the Transform first increases every MS band by the high-goal Pan band and afterward partitions every item by the amount of the MS groups. The SVR and RE strategies are comparable, yet include more complex estimations for the MS aggregate for better combination quality.

- A high-goal Pan picture is first fragmented into a group of low-goal Pan pictures using wavelet coefficients (spatial subtleties) for each level in wavelet combination. Individual MS picture groups then take the place of the low-goal Pan at the goal level of the first MS picture. By playing out an opposite wavelet change on every MS rally with the corresponding wavelet coefficients, high-goal spatial detail is introduced into every MS band.

C. *Limitations of Existing Fusion Techniques:*

Many exploration papers have announced the limits of existing combination methods. Shade mutilation is the main concern. Another common difficulty is that the quality of the combination is frequently dependent on the administrator's combined experience and the intertwining of the informational index. There has never been a programmed arrangement that consistently delivers outstanding combinations for various informational resources.

To diminish the shading twisting and further develop the combination quality, a wide assortment of procedures have been grown, every particular to a specific combination method or picture sets. For instance:

- For IHS combination, a typical procedure is to match the Pan to the I band before the substitution, stretch the H and S groups before the converse IHS change, or stretch individual I, H, or S groups concerning individual informational indexes.
- In PCA combination, recommended arrangements have been, for instance, extending the foremost parts to give a round circulation, or disposing of the main head part.
- With math blend techniques, shading twisting fluctuates relying on the band mixes being intertwined. Pre-processing and the administrator's combination experience are essential to accomplishing a decent combination result.
- In wavelet combination, numerous variations of wavelet combination have been created to manage shading contortion issues.

By choosing a legitimate combination method and applying a suitable change procedure, victories can be accomplished for the combination of SPOT Pan or IRS Pan pictures with other low-goal MS pictures, such as Landsat TM, SPOT MS, or IRS MS. Be that as it may, the administrator's experience assumes a significant part in the achievement.

7.4 QUALITY EVALUATION

1. Figure 7.1 shows one illustration of the intertwined WorldView-2 regular shading picture when the dish honing. Because as far as possible, the skillet honing consequences of other otherworldly groups and other satellite pictures can't be introduced here.
2. A professional visual picture investigation was performed for the combination quality assessment. Quantitative assessment was not utilized because writing surveys and our past assessment tests exhibited that no current quantitative techniques can give reliable and persuading assessment results [36].
3. To stay away from predisposition in the visual examination,
   1. every one of the pictures when dish honing was shown under a similar representation condition (i.e., a similar picture region was shown and a similar histogram extending was applied to every one of the pictures); and
   2. free experts were welcome to break down and look at the when container honing pictures at a 1:1 or bigger scope without showing the names of the skillet honing strategies. The experts associated with the assessment incorporate picture combination specialists, picture combination administrators, remote detecting, and satellite picture clients. Practically consistent assessment results were gotten, which are summed up in the end underneath.
4. Before perusing the assessment finishes of the experts in question, peruses can break down and think about the when skillet honing pictures in Figure 7.1 to give their assessment remarks as far as picture combination quality.
5. As shown in Figure 7.1, ERDAS' Resolution Merge (Principal Components technique) and ENVI's PC Spectral Sharpening give similar dish honing results in terms of shading and sharpness. ENVI's Gram Schmidt Spectral Sharpening mutilates shading in the same way that Resolution Merge and PC Spectral Sharpening do, but with more sharpness. These findings support the hypothesis that Gram change, such as head part change, is a measurable methodology.

7.5 RESULTS

The proposed combination execution portrayals technique is demonstrated by the evaluation of a broad range of multiresolution and multiscale picture combination calculations, which have long been believed to be among the best procedures for blending two images [1–6].

By repeated aim lowering and extraction of high-recurrence portions into pyramid sub-groups, input images are first decomposed into an alleged pyramid portrayal. The upsides of the information pyramids are then used

Table 7.1 Multiresolution combination approaches

Fusion Scheme	$Q^{AB/F}$	Q_c	Q_A	$L^{AB/F}$	$N^{AB/F}$
Averaging	0.495	0.384	0.205	0.651	0
Contrast	0.620	0.233	0.466	0.30	0.210
Laplacian	0.720	0.215	0.475	0.326	0.351
Gradient	0.540	0.427	0.370	0.60	0.030
DWT mres	0.564	0.191	0.495	0.333	0.391
DWF MSc	0.683	0.286	0.537	0.424	0.149
BMI	0.592	0.286	0.387	0.557	0.333

to create a new picture pyramid using a pre-defined combination standard. As a result, elements of various scales can be considered separately, regardless of whether they have covering regions in the image [1,2].

A few multiresolution depictions also include direction awareness and data isolation based on the direction [3–6]. Multiresolution reproduction is used to create the fused picture when a new, interconnected pyramid is created. The multiscale technique is similar, but it does not use target reduction, resulting in an unfathomably repeating pyramid representation [3,7]. There were a few multiresolution combination approaches that were considered as shown in Table 7.1.

7.6 FUTURE TRENDS

Albeit different picture combinations and objective execution assessment techniques have been proposed, as of now, there are as yet many open-finished issues in various applications. In this part, the future patterns in various application spaces, i.e., remote detecting, clinical conclusion, reconnaissance, and photography, are dissected.

1. In remote detecting, the serious issue is how to diminish visual mutilations while combining multispectral, hyperspectral, and panchromatic pictures. Besides, albeit the information pictures are generally caught with similar stage, distinctive imaging sensors don't by and large zero in on a similar course, and their securing minutes are likewise not unequivocally indistinguishable.
2. In the present circumstance, exact enlistment is trying because of the huge goal and phantom distinction among the source pictures. Finally, because of the quick improvement of remote detecting sensors, creating novel calculations for the combination of pictures caught by clever airplane or satellite sensors will be a hot examination subject.
3. In the clinical conclusion field, exact enlistment is additionally a difficult theme since the clinical pictures are normally caught with various

securing modalities. All the more significantly, planning strategies focusing on explicit clinical issues is a difficult and nontrivial undertaking of clinical analysis. The explanation is that the plan of combination techniques requires both clinical area information and algorithmic bits of knowledge. Additionally, since the ideal combination exhibitions perhaps not be equivalent to those of general picture combination, the true assessment of the clinical picture combination techniques is likewise a difficult theme.

4. In observation applications, two significant necessities ought to be thought of. On one hand, combination techniques are relied upon to be computationally proficient, to be utilized for ongoing reconnaissance applications. Then again, since the picture procurement conditions might differ drastically in the open-air climate, planning combination strategies hearty to defective conditions such as clamour, under or over-openness is a vital theme.

5. In the photography field, since the information pictures are caught 100% of the time on various occasions, moving items might show up at various areas during the catching system, and in this manner, produce apparition antiquities in the melded picture. To tackle this issue, planning techniques uncaring toward such misregistration is a difficult subject around here. Besides, another examination heading is how to apply picture combination calculations into implanted customer camera frameworks for commonsense photography applications.

7.7 CONCLUSION

High-goal and multispectral remote detecting pictures are a significant information hotspot for getting the huge scope and point-by-point geospatial data for an assortment of utilizations. Visual translation, advanced grouping, and shading picture perception are three methodologies frequently used to get or show point-by-point geospatial data. Contingent on the reason for a given application, (1) a few clients might want a combination result that shows more detail in shading, for better picture understanding or planning; (2) some might want a combination result that works on the precision of computerized arrangement; and (3) some others might want an outwardly lovely intertwined shading picture, exclusively for perception purposes. In this manner, particular strategies for planning focused combination, order arranged combination, and perception situated picture are popular. Presently, picture combination (or container honing) strategies have been demonstrated to be successful devices for giving better picture data to visual translation, picture planning, and picture-based GIS applications. The insights-based combination method can meld Pan and MS pictures of the new satellites, just as pictures of SPOT and IRS, bringing about limited shading mutilation, expanded detail, and normal tone and element coordination (see title page and Figure 7.4). In any case, there is as yet an absence of viable methods for order arranged combination and representation situated combination. An underlying programmed procedure for perception arranged combination, called shading improved combination, has been created.

ACKNOWLEDGEMENT

We would like to thank the support of Chandigarh Group of Colleges Landran Branch College of Engineering (COE), Department of Computer Science and Engineering, and Chandigarh Engineering College (CEC), Department of Electronics and Communication Engineering. It is with great appreciation that I thank CCG to give me such a wonderful guide for my research publication too.

REFERENCES

[1] Toet, A., van Ruyven, L., & Velaton, J. (1989). Merging thermal and visual images by a contrast pyramid. *Optical Engineering, 28*(7), 789–792.

[2] Zhang, Z., & Blum, R. (1999). A categorization of multiscale decomposition-based image fusion schemes with a performance study for a digital camera application. *Proceedings of the IEEE, 87*(8), 1315–1326.

[3] Li, H., Munjanath, B., & Mitra, S. (1995). Multisensor image fusion using the wavelet transform. *Graphical Models and Image Proceedings, 57*(3), 235–245.

[4] Chibani, Y., & Houacine, A. (2003). Redundant versus orthogonal wavelet decomposition for multisensor image fusion. *Pattern Recognition, 36*, 879–887.

[5] Chipman, L., Orr, T., & Graham, L. (1995). Wavelets and image fusion. *Proceedings of SPIE, 2569*, 208–219.

[6] Burt, P., & Kolczynski, R. (1993). Enhanced image capture through fusion. In *Proceedings of 4th International Conference on Computer Vision* (pp. 173–182), Berlin.

[7] Petroviü, V., & Xydeas, C. (October 1999). Computationally efficient pixellevel image fusion. In *Proceedings of Eurofusion99*, Stratford-upon-Avon (pp. 177–184).

[8] Xydeas, C., & Petroviü, V. (February 2000). Objective image fusion performance measure. *Electronics Letters, 36*(4), 308–309.

[9] Qu, G., Zhang, D., & Yan, P. Information measure for the performance of image fusion. *Electronics Letters, 38*(7), 313–315.

[10] Piella, G., & Heijmans, H. (2003). A new quality metric for image fusion. *Proceedings International Conference on Image Processing, 3*, 173–176.

[11] Petroviü, V., & Xydeas, C. (2004). Evaluation of image fusion performance with visible differences. *Proceedings of ECCV, Prague, LNCS 3023*, 380.

[12] Burt, P., & Adelson, E. The Laplacian pyramid as a compact image code. *IEEE Transactions on Communications, COM, 31*(4), 532–535.

[13] Zhang, Y. (2004). Understanding image fusion. *Photogrammetric Engineering and Remote Sensing*, 657–661.

[14] Mandic, D. P., Obradovic, D., Kuh, A., Adali, T., Trutschel, U., Golz, M., De Wilde, P., Barria, J., Constantinides, A., & Chambers, J. (2005). Data fusion for modern engineering applications: an overview. *International Conference on Artificial Neural Networks 2005, LNCS*, 3697, 715–721.

[15] Qu, G., Zhang, D., & Yan, P. (2001). Medical image fusion by wavelet transform modulus maxima. *Optics Express, 9*(4), 184–190.

[16] Van Dun, B., Wouters, J., & Moonen, M. (2007). Improving auditory steady-state response detection using independent component analysis on multichannel EEG data. *IEEE Transactions on Biomedical Engineering, 54*(7), 1220–1230.

[17] Kwak, K. C., & Pedrycz, W. (2007). Face recognition using enhanced independent component analysis approach. *IEEE Transactions on Neural Networks, 18*(4), 530–541.

[18] Blum, R. S., Xue, Z., & Zhang, Z. (2005). An overview of image fusion. In R. S. Blum (ed.), *Multi-Sensor Image Fusion and its Applications*. Marcel Dekker.

[19] Nikolov, S., Hill, P. R., Bull, D. R., & Canagarajah, C. N. (2001). Wavelets for image fusion. In A. Petrosian & F. Meyer (eds.), *Wavelets in Signal and Image Analysis, from Theory to Practice*. Kluwer Academic Publishers.

[20] Sadjadi, F. (2005). Comparative image fusion analysis. In *IEEE International Workshop on Object Tracking and Classification in and Beyond the Visible Spectrum*. USA.

[21] Burt, P. J., & Adelson, E. H. (1983). The Laplacian pyramid as a compact image code. *IEEE Transaction on Communications, 31*(4), 532–540.

[22] Zeng, J., Sayedelahl, A., Gilmore, T., & Chouikha, M. (2006). Review of image fusion algorithms for unconstrained outdoor scenes. *Proceedings of International Conference on Signal Processing*, 2.

[23] Rogers, S. K., Tong, C. W., Kabrisky, M., & Mills, J. P. (1989). Multisensor fusion of radar and passive infrared imagery for target segmentation. *Optical Engineering, 28*(8), 881–886.

[24] Huntsberger, T., & Jawerth, B. (1993). Wavelet-based sensor fusion. *Proceedings of SPIE, 2059*, 488–498.

[25] Li, H., Manjunath, B. S., & Mitra, S. K. (1995). Multi-sensor image fusion using the wavelet transform. *Graphical Models and Image Processing, 57*(3), 235–245.

[26] Chipman, L. J., Orr, Y. M., & Graham, L. N. (1995). Wavelets and image fusion. In *Proceedings of International Conference on Image Processing* (pp. 248–251), USA.

[27] Rockinger, O. (1997). Image sequence fusion using a shift-invariant wavelet transform. In *Proceedings of International Conference on Image Processing* (pp. 288–291). Singapore.

[28] Chibani, Y., & Houacine, A. (2000). On the use of the redundant wavelet transform for multisensory image fusion. In *Proceedings of IEEE International Conference on Electronics, Circuit, and Systems* (pp. 442–445). USA.

[29] Waxman, A. M., Fay, D. A., Gove, A. N., Siebert, M., Racamoto, J. P., Carrick, J. E., & Savoye, E. D. (1995). Color night vision: Fusion of intensified visible and thermal IR imagery. *Proceedings of SPIE Conference on Synthetic Vision for Vehicle Guidance and Control, 2463,* 58–68.

[30] Sharma, R. K., Leen, T. K., & Pavel, M. (1999). Probabilistic image sensor fusion. *Advances in Neural Information Processing Systems 11.* The MIT Press.

[31] Pohl, C., & van Genderen, J. L. (1998). Multisensor image fusion in remote sensing: concepts, methods and applications. *International Journal of Remote Sensing, 19(5),* 823854.

[32] Pattichis, C. S., Pattichis, M. S., & Micheli-Tzanakou, E. (2001). Medical imaging fusion application: an overview. *Asilomar Conference on Signals, Systems and Computers,* 1263–1267.

[33] Cvejic, N., Bull, D., & Canagarajah, N. (2007). Improving fusion of surveillance images in sensor networks using independent component analysis. *IEEE Transactions on Consumer Electronics, 53(3),* 1029–1035.

[34] Omar, Z., Mitianoudis, N., & Stathaki, T. (2010). Two-dimensional Chebyshev polynomials for image fusion. In *28th Picture Coding Symposium,* (pp. 426–429), Japan.

[35] Loza, A. (22 June 2007). Statistical model-based fusion of noisy images. *Computer Science Departmental Seminar,* University of Bristol.

[36] Zhang, Y. (2008). Methods for image fusion quality assessment – a review, comparison, and analysis. *The XXI ISPRS Congress, Beijing, July 3–11, 2008. The International Archives of the Photogrammetry, Remote Sensing and Spatial Information Sciences,* XXXVII (Part B7), 1101.

Chapter 8

Meta-Heuristic Algorithms for Optimization

K. Thippeswamy

CONTENTS

8.1 INTRODUCTION

Examples of meta-heuristics include particle swarm optimization (PSO), ant colony optimization (ACO) algorithms, etc., which are used in many areas of optimization. Many such meta-heuristic algorithms have been used to solve optimization problems in industrial and engineering applications. The following sections give details of the reviewed meta-heuristics [1–5].

DOI: 10.1201/9781003355960-8

8.1.1 Bat Algorithm

Yang developed the bat algorithm which is bio-inspired. It is based on the feature of the echolocation behaviour of micro-bats. Parameters such as frequency tuning and automatic zooming will help in obtaining a range of solutions in the population which is mimicked by varying the pulse rate and loudness of bats in hunting the prey.

8.2 BASIC CONCEPTS

For detecting the prey, micro-bats emit sonar waves called echolocation. Bats know the actual difference between the prey and obstacles from the reflected waves. The special ability of echolocation help micro-bats in identifying their prey even in darkness. Pulse rates vary depending on the type of species and can be related to hunting strategy. Micro-bats use a frequency range from 25 to 150 kHz with a pulse rate of 200 pulses per second and with a velocity of 350 m/s. These wavelengths are the same as that of their prey size.

8.2.1 Pseudo Code of Bat Algorithm

Based on the above characteristics and description of bat echolocation, Xin-She Yang developed the bat algorithm with the following three idealized rules

1. All bats use echolocation to sense distance, and they also "know" the difference between background and prey barriers in some magical way.
2. Bats fly randomly with velocity v_i at position x_i with a frequency f and loudness A_0 to search for prey. They can automatically adjust the frequency of their emitted pulses and adjust the rate of pulse emission $r \in [0; 1]$, depending on the proximity of their target.
3. Although the loudness can vary in many ways, we assume that the loudness varies from a large A_0 to a minimum constant value A_{min} [6–10].

A mathematical model in the design of the bat algorithm is discussed next.

Every bat i is associated with a velocity v_i^t and a location x_i^t, with t as iteration, in a search space. Among all the bats, there exists a current best solution x_*. Mathematical formula for all three rules can be interpreted as follows:

$$f_i = f_{min} + \beta (f_{max} - f_{min}) \tag{8.1}$$

$$v_i^t = v_i^{t-1} + f_i(x_i^{t-1} - x_{\text{cgbest}}) \tag{8.2}$$

$$x_i^t = x_i^{t-1} + v_i^t \tag{8.3}$$

where β is a uniform distribution $\in [0, 1]$. $f_{\min}$ and $f_{\max}$ are taken to be 0 and 1, respectively.

Updating the new location is as in 8.4

$$x_{\text{new}} = x_{\text{old}} + \epsilon A^t \tag{8.4}$$

where x_{new} is the new position value of the bat algorithm, x_{old} is the previous position of the bat algorithm, ϵ is a random number between $[-1, 1]$, and A is the loudness.

As the iteration advances, the loudness A_i and the pulse emission rate r is also updated accordingly. When the bat approaches the prey, the pulse rate increases and the loudness decreases and is evaluated as given in 8.5 and 8.6

$$A_i^{t+1} = \alpha \, A_i^t \tag{8.5}$$

$$r_i^{t+1} = r_i^0 [1 - \exp(-\gamma t)] \tag{8.6}$$

where α and γ are constants. The values are set to 0.9. We have $A_i^t \longrightarrow 0$ and $r_i^t \longrightarrow r_i^0$ as $t \longrightarrow \infty$ (Table 8.1).

8.2.2 Pseudocode of Bat Algorithm

Pseudocode of original bat algorithm.

Table 8.1 Configuration parameters

Parameters	Definition	Value
Pop size	Size of population	25
Number of iterations	Maximum number of iterations	1,000
$f_{\max}$	Maximum frequency	2
$f_{\min}$	Minimum frequency	0.25
r_i	Pulse rate	0.75
A_i	Loudness of emission	0.5
γ	Increment value of pulse rate	0.9
α	Decrement value of loudness	0.9

Objective function f(x), x = (x₁, x₂, ... , x_d)ᵀ

Objective function $f(x)$, $x = (x_1, x_2, \ldots , x_d)^T$
Initialize the bat population x_i and v_i
Define pulse frequency f_i at x_i
Initialize pulse rate r_i and the loudness A_i
while *($t <$ max number of iterations)*
Generate new solutions by adjusting frequency,
esandositions by equations (8.1) to (8.3)
if *(rand $> r_i$)*
Select a solution among the best solutions, generate a local solution around the selected best solution
end if
Generate a new solution by flying randomly
(rand $< A_i$ and $f(x_i) < f(x^*)$)
Accept the new solutions
Increase r_i and reduce A_i
end if
Rank the bats and find the current best x^*

8.3 ANT COLONY OPTIMIZATION

Marco Dorigo proposed an ACO algorithm based on the behaviour of ants looking for a path between their colony and a source of food. ACO is an actual popular concept for combinatorial optimization problems.

8.3.1 ACO in Brief

Ants cannot find the shortest path between their colony to which they belong and the source of food as they do have sight. The decision of choosing the path in the beginning is random. Ants communicate by a chemical substance called pheromone which they leave as they travel. As more number of ants travel on the same path, the pheromone concentration increases. Pheromone slowly evaporates with time. Ants follow the path that has concentrated pheromones.

8.3.2 Standard ACO

The probability of ant k to choose the jth position from the ith position during iteration is given by the following equation:

$$p_{ij}^k(t) = \begin{cases} \dfrac{[\tau_{ij}(t)]^\alpha [\eta_{ij}]^\beta}{\sum_{k \in \text{allowed}_k}[\tau_{ik}(t)]^\alpha [\eta_{ik}]^\beta} & \text{if } j \in \text{allowed}_k \\ 0 & \text{otherwise} \end{cases} \tag{8.7}$$

The amount of pheromone at each position is as follows:

$$\tau_{ij}(t + 1) = (1 - \rho)\tau_{ij}(t) + \rho\Delta\tau_{ij}^{best}(t)$$

8.3.3 Parameters of ACO

α and β are used in the probabilistic decision rule.
α is the parameter to regulate the influence of pheromone.
β is the parameter to regulate the influence of η_{ij}
ρ is the evaporation parameter.

When $\rho = 1$, there is full evaporation, and when $\rho = 0$, there is no evaporation. Hence, the value is set to 0.5, and the ant colony size is 50.

8.3.4 Algorithm of ACO

Step 1: For every ant, evaluate the value of probability.
Step 2: Update pheromone.
Step 3: Repeat the steps till the set number of iterations.

8.4 PSO ALGORITHM

PSO algorithm was proposed by Kennedy and Eberhart in 1995. The algorithm is a population-based stochastic optimization algorithm. The algorithm is based on the social behaviour of bird flocking or fish schooling where each fish or bird is considered a particle. These particles fly with a certain velocity to find the global best g_{best} solution after traversing through several local best solutions in each iteration. It has been found highly effective in solving several optimization problems.

8.4.1 PSO Basics

X_i denotes the position of a particle i; V is the velocity of particle i moving at time t; P_i is the personal best position; g_{best} (g) represents global best; and k is the dimension of the problem space. Figure 8.1 depicts the new position of the particle using the PSO method. The particles are updated according to the following equations:
Updating the particle position

$$X_i(t + 1) = X_i(t) + V_i(t + 1)$$

$$V_i(t + 1) = W * V_i(t) + c_1(P_i(t) - X_i(t) + c_2(g(t) - X_i(t))$$

Updating velocity in time steps t, jth component of ith particle.

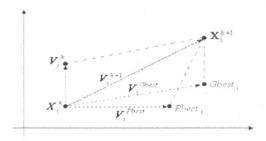

Figure 8.1 Examining new position in PSO.

$$V_{ij}\,(t+1)= w\ V_{ij}(t) + r_1c_1\,(p_{ij}(t)-X_{ij}(t))+ r_2c_2\,(g_i(t)-X_{ij}(t))$$

where the first term is the inertia term, the second term is called the cognitive component, and the third term is called the social component.
r_1 and r_2 are uniformly distributed $(0, 1)$.
c_1 and c_2 are called acceleration coefficients.
Equation for updating position of particle for *j*th component is as follows:

$$X_{ij}\,(t+1)= X_{ij}(t) + V_{ij}(t+1)$$

8.4.2 PSO Algorithm

For every particle, initialize position, velocity, and cost and then generate random solutions
Update personal best, and global best for every particle.
Calculate the fitness function for each particle.
Evaluate velocity, position, and personal best for all particles using the above-mentioned equations.
Find the personal best by comparing particle cost, position with particle best cost, and particle best position, respectively update global best
Repeat the above steps till the end of iterations, store the best value, and iterate till the maximum number of iterations.

8.4.3 Parameters of PSO Algorithm

Upper bound of the search space = maximum of search space
Lower bound of search scope = minimum of search space range
Acceleration coefficients:
Personal acceleration coefficient $c_1 = 2$
Social acceleration coefficient $c_2 = 2$
Inertia coefficient $w = 1$
Population size = 50
Maximum iterations = 1,000

Table 8.2 Description of datasets

Dataset	Dimension type	Total no. of instances	No. of attributes	No. of classes
Breast cancer Wisconsin	Integer	683	9	2
Zoo	Integer	101	17	
Wine	Integer, real	150	13	3

8.5 EXPERIMENTAL ANALYSIS

The proposed research work is designated to perform comparative analysis on well-known datasets in order to test the efficiency of the algorithm, i.e., PSO, bat, and ACO algorithms.

8.5.1 Dataset Description

Datasets are selected from the datasets of UCI available at http://archive.ics. uci.edu/ml/datasets.html these datasets are namely Breast Cancer Wisconsin, Wine, and Zoo. The detailed descriptions of the datasets are given in Table 8.2.

8.6 RESULT ANALYSIS

Performance is measured on the basis of means, standard deviation, normalized absolute error (NAE), and accuracy. The arithmetic mean is calculated to measure the central tendency of data. Figure 8.2 shows the calculated weighted arithmetic mean values by using three optimization algorithms, i.e., bat, ACO, and PSO algorithms.

The weighted arithmetic mean for a given dataset is as follows:

$$\frac{\sum_{i=1}^{n} w_i X_i}{\sum_{i=1}^{n} w_i}$$

where w_i represents the *weight* associated with element X_i; it is equal to the number of times that the element appears in the dataset. Figure 8.2 weighted arithmetic mean of three algorithms.

Standard deviation (σ) shows how much variation or dispersion from the average exists. Standard deviation is calculated to measure the dispersion of data. Figure 8.3 shows the calculated standard deviation values by using three optimizations, i.e., bat, ACO, and PSO algorithms.

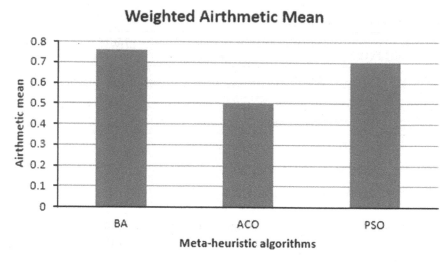

Figure 8.2 Weighted arithmetic mean.

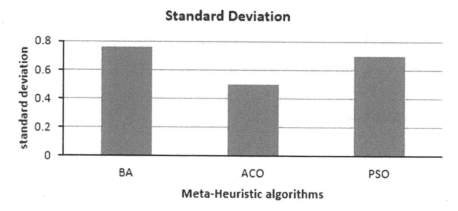

Figure 8.3 Standard deviation comparisons.

Normalization is the process of isolating the statistical error in repeated measured data. These ratios only make sense for ratio measurements in terms of levels of measurement. Figure 8.4 shows the calculated NAE values by using three optimization algorithms, i.e., bat, ACO, and PSO algorithms.

Classification accuracy is defined as follows:

$$\text{Accuracy} = \frac{\text{Number of classified objects}}{\text{Number of objects in datasets}} * 100\%$$

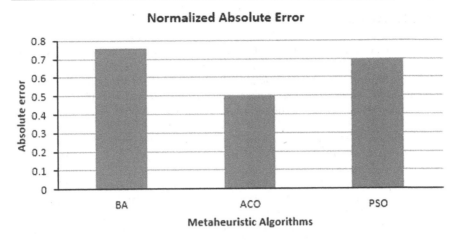

Figure 8.4 Comparison of algorithms in normalized absolute error.

Table 8.3 Accuracy of algorithms

Datasets	PSO	ACO	Bat algorithm
Wine	93.78	95.45	95.12
Zoo	95.480	96.06	96.78
WBC	95.150	96.13	96.25

The accuracy of classification of the three meta-heuristic algorithms evaluated is given in Table 8.3 in percentage. From the results, we see that the bat algorithm yielded the highest accuracy in all the standard datasets.

8.7 CONCLUSION

The three algorithms belong to population-based methods. According to this method, a set of solutions will be improvised at each iteration. The bat algorithm is found to be a successful optimization algorithm. The three algorithms were tested using the standard datasets from UCI. The performance of these algorithms is analyzed by evaluating weighted arithmetic mean, NAE, and standard deviation, and accuracy is considered for comparison purpose. The bat algorithm is found to be a successful optimization algorithm. The success of the bat algorithm is due to features of frequency tuning (updating the value of frequency from f_{min} to f_{max}), automatic zooming (switching to an area where there are chances of getting hopeful solutions), and parameter control (changing the value of loudness and pulse rate at each iteration).

REFERENCES

[1] Martinez, A. W., Phillips, S. T., Wiley, B. J., Gupta, M., & Whitesides, G. M. (December 2008). FLASH: a rapid method for prototyping paper-based microfluidic devices. *Lab on a Chip*, 8(12), 2146–2150. 10.1039/b811135a

[2] Pollock, N. R., et al. (September 2012). A paper-based multiplexed transaminase test for low-cost, point-of-care liver function testing. *Science Translation Medicine*, 4(152), 152ra129. 10.1126/scitranslmed.3003981

[3] Nilsson, I., Johnson, A. E., & von Heijne, G. (August 2003). How hydrophobic is alanine? *The Journal of Biological Chemistry*, 278(32), 29389–29393. 10.1074/jbc.M212310200

[4] The amino acid alanine – benefits, information on supplements, articles, links, news, advice. http://www.vitaminstuff.com/amino-acid-alanine.html (accessed March 14, 2022).

[5] Babson, A. L., Shapiro, P. O., Williams, P. A. R., & Phillips, G. E. (March 1962). The use of a diazonium salt for the determination of glutamic-oxalacetic transaminase in serum. *Clinica Chimica Acta*, 7(2), 199–205. 10.1016/0009-8981(62)90010-4

[6] Lippi, U., & Guidi, G. (June 1970). A new colorimetric ultramicromethod for serum glutamicoxalacetic and glutamic-pyruvic transaminase determination. *Clinica Chimica Acta*, 28(3), 431–437. 10.1016/0009-8981(70)90069-0

[7] Canepari, S., Carunchio, V., Girelli, A. M., & Messina, A. (June 1994). Determination of aspartate aminotransferase activity by high-performance liquid chromatography. *Journal of Chromatography B: Biomedical Sciences and Application*, 656(1), 191–195. 10.1016/0378-4347(94)00069-7

[8] Karmen, A. (January 1955). A note on the spectrometric assay of glutamic-oxalacetic transaminase in human blood serum. *The Journal of Clinical Investigation*, 34(1), 131–133.

[9] Itoh, H., & Srere, P. A. (June 1970). A new assay for glutamate-oxaloacetate transaminase. *Analytical Biochemistry*, 35(2), 405–410. 10.1016/0003-2697(70)90202-2

[10] Janasek, D., & Spohn, U. (February 1999). Chemiluminometric flow injection analysis procedures for the enzymatic determination of L-alanine, α-ketoglutarate and L-glutamate. *Biosensors and Bioelectronics*, 14(2), 123–129. 10.1016/S0956-5663(98)00115-8

Chapter 9

Internet of Medical Things Devices: A Review

Sachin Dhawan, Arun Kumar Rana, Sumit Kumar Rana, and Kashif Nisar

CONTENTS

9.1 INTRODUCTION

A single sensor or numerous sensors may be used in Internet of Things (IoT) devices such as wearable or stand-alone devices are possible [1]. Patients and physicians can both wear these gadgets, which allow them to monitor their patients remotely. Some of these flaws can be mitigated to a large extent with IoT-based healthcare solutions. We have included an alert mechanism in this system to notify the doctor when a health parameter surpasses a threshold. Every few seconds, the patient's condition will be monitored, and fresh data will be updated. As a result, patients can be

DOI: 10.1201/9781003355960-9

watched in any emergency situation. Hence, the system will be a portable remote monitoring system with a power source.

Patients may be treated anywhere in the world thanks to IoT healthcare monitoring, which is a fantastic choice for underdeveloped nations. A heartbeat detection method, a temperature detection mechanism, and a panic warning mechanism are all included in the system. This device may be used by a caregiver, health expert, or nursing staff to remotely monitor critical health parameters of patients or people in need. We used an Arduino Uno as the system's microcontroller, a temperature sensor to calculate the patient's body temperature, and a pulse sensor to calculate the patient's heart rate. We also used an ESP8266 to connect the hardware to the Internet and provide WiFi so that the patient's vitals could be sent to the cloud.

The patient's vitals are presented on the system's liquid crystal display (LCD), and the data are displayed on the ThingSpeak cloud, which is an IoT platform. The outcomes are represented graphically. These numbers are changed every few seconds, allowing the patient's vitals to be updated on a regular basis. When the patient's vitals reach a specified level, an email alert is sent to the nursing staff, doctor, or carer using the If This Then That (IFTTT) service. We've also installed an alert button so that if a patient becomes uneasy or ill, he may push the emergency button, which will activate a panic alarm, alerting the patient's caregiver or nursing staff, who can then assess the situation. Any casualties can be avoided as a result of this. This system has the potential to be highly useful in the healthcare infrastructure, and it may be expanded to include many devices and various sensors in the future. Artificial intelligence and machine learning may be applied in the future to improve the accuracy and efficiency of this system. Predictions regarding the patient's general health can be made, and additional analysis of their health data can be done. This type of approach can minimize not just the cost of therapy but also the amount of labour that nurses have to do. It has the potential to be extremely valuable in the remote monitoring of a patient's health.

9.2 LITERATURE REVIEW

Jaiswal et al. [2] devised an automated system that collects patient statistics from medical equipment equipped with sensors and then processes the data before uploading it to the medical centre's cloud. Docker containers were also utilized for storage, distribution, and other purposes. Data from sensor nodes are gathered, encoded, and transmitted to a server through a wireless channel using the appropriate software. Before being pushed to the server, data (patient's temperature, blood pressure, etc.) are entered into the Raspberry Pi. For further processing of acquired data, a Docker container and a local database on the server are utilized, and the data are sent to doctors, nursing staff, and hospitals to monitor and diagnose health problems. The docker container users and the rest of the components employ proper authentication and data encryption. Apart from data storage, the system also provides services such as obtaining information from physicians, users, and hospitals. An android app developed in [3] monitors a patient's temperature and pulse rate, which are basic indicators of a change in the patient's health. These sensors feed data to a Raspberry Pi, and the data are subsequently uploaded to the cloud. The ds18b20 is used to measure temperature. Temperature and heart rate are constantly monitored by the Raspberry Pi. The sensor 11574 is used to measure heart rate. For continuous monitoring, these measured values are transferred to the cloud (ThingSpeak). These data may also be accessed by the doctor or carer, who can then take necessary action if there are any discrepancies. This information is available on the Android App, and a doctor or caretaker may use it to take any additional activities that are necessary based on the observations. In the event of an emergency, they can request assistance through the app by dialling an emergency number. There is a feature that allows you to send text messages to the doctor, such as "patient is not well, needs aid." Another health monitoring system [4] employs three non-contact sensors to detect heart rate, respiration rate, SpO_2, and temperature in that order. The body sensor network employs signal processing and data transmission modules that allow the gathered data to be wirelessly sent over the Internet. The Arduino Uno receives the sensor input. The oxygen level in the blood is measured by SpO_2. It's defined as the proportion of oxygenated haemoglobin to non-oxygenated haemoglobin in the blood (red blood cells). The SpO_2 levels alter when the intensity of red light changes. For a healthy person, the usual range is between 95% and 100%. The temperature is monitored using a temperature sensor from the LM series.

In paper [5], Saranya and Maheswaran propose a data mining smartphone application. Sensors are employed once again to create a health illness prediction and diagnostic system. Using a wireless connection, this technology measures several patient parameters such as pressure, heartbeat, and temperature. In this project, Arduino is once again connected to a variety of sensors. The participant wears these three wearable sensors (1, 2, and 3), which monitor the heart rate, body temperature, and pressure level of the human body, accordingly. On the LCD, these related measures will be presented. These variables are utilized to intelligently anticipate the patient's illness. The patient is considered healthy if all three of the detected parameters are within the threshold levels.

Patient data are saved in the cloud, allowing healthcare providers to monitor their patients at any time from a remote location. Wearable sensor nodes that analyze time alarm messages and send them to concerned cellphones capture deviations in data values. The data mining smartphone application, in addition to monitoring, disease prediction, and diagnosis, may be used with this system.

In paper [6], Sambhram has built another Raspberry Pi-based health monitoring system that uses wireless connections and has minimal power consumption while providing maximum capability with many sensors and user mobility. Using a Raspberry-Pi module, data are gathered from ECG (electrocardiogram), BP (blood pressure), heart rate, and temperature sensors, which are then analyzed and a judgement is made whether the readings read from the patient are normal or pathological using the Fog computing approach. These are transferred to cloud storage and may be accessed by connecting to the Internet on a smartphone. In the event that aberrant values are read, a facility for SMS notification, message, and other messages will be delivered to the user's smartphone, allowing him or her to take care of the patient in severe emergency scenarios. This paper [7] devised an automatic system to monitor a patient's body temperature, heart rate, bodily motions, and blood pressure, as well as anticipate if the patient has a chronic ailment or disease based on the numerous health metrics and many other symptoms. In level one, data are captured and saved on the cloud using vibration sensors, BP sensors, and pulse and temperature sensors. To obtain usable information, it is further filtered and categorized. Data mining techniques are used to advance the analysis or prediction. The Raspberry Pi takes the data from various vital signs of patients and updates it in a MySQL database. In the event that vitals deviate from threshold levels, this is utilized to send information to concerned doctors, caregivers, and others through email, messages, and other means. Aside from the data acquired via sensors, patients were also questioned about their symptoms, and these data were compared with current information to forecast disease/disorder, making it an effective Expert system.

The same three sensors are used to detect body temperature, pulse rate, and heartbeat in [8]. The ECG sensor delivers signals to the Arduino Mega microcontroller, which is connected to the Internet through a WiFi module. This information is sent to the Internet app ThingSpeak by Arduino. The patient's heart rate was 121 beats per minute (BPM), and he was monitored with an IBI of 1,826 ms. The data were collected when the subject was both calm and eager. The heart rate was 80–90 BPM in a calm position and 120 BPM when stressed or anxious. When a patient is stressed, his or her temperature rises, and the temperature of the environment also contributes to the rise in temperature. When the patient lies down to rest, his or her body temperature reduces to its ideal level.

With the objective of having an effective communication between patient and doctor, another healthcare monitoring system is suggested in [9]. To provide better healthcare facilities in rural areas, basic health checkups, a more technology-oriented system is developed. Such an intelligent system will have a facility of some patient accounts that will maintain patient data collected using some hardware. These data can be read by patients also through an app on their mobiles using Bluetooth module HC-05 and the same data will also be uploaded to the fusion table's patient database at the same time.

Another health monitoring system made using an Arduino ESP32, a heart rate and temperature sensor, Android Studio, Firebase, and Thingspeak [10] employ an Android application with a portable measurement device. The DS18B20 and pulse sensors are attached to the Arduino ESP32 microcontroller board. Temperature and heart rate values are entered via sensors, displayed on an LCD, and delivered in real-time via WiFi to the ThingSpeak platform.

Data from the past is kept on the platform. In the event that readings deviate from the required values, the system delivers a real-time message via the Line application. The Firebase Real-time Database stores information about registered members such as their identity, heart rate, temperature, and measurement equipment, among other things. The IoThNet topology is described by the authors in [11], which is a network of multiple sensors, computation, and storage capabilities combined in a heterogeneous system. It connects mobiles, servers, laptops, optical instruments, and other devices to measure ECG, BP, SpO_2, and temperature. Apple Watch uses a huge number of sensors based on infrared and visible light to monitor calorie count, distance, step count, and other metrics. The NFC antenna connects two devices, a wearable and a mobile phone, to establish communication.

IoT-based healthcare system has also been applied to determine cases of heatstroke, hypothermia, bradycardia, and tachycardia in [12]. The system uses Raspberry Pi along with DS18B20, ADC1015, ADXL345, and a heartbeat sensor. The use of ATmega328 can also be seen in [13] and [14,15] to measure the temperature of patients, alerting the caretaker/doctor along with pulse sensor data and pushing data to cloud. Similar work is also achieved using ESP8266 in [16]. More specific health monitoring system with IoT is seen with BAN (body area network) developed in [17] to collect information such as heat, circulatory strain, and pulse rate of patients and observe these parameters using a specialized application.

In [18], the authors have used (TI) MSP430F5529 which is a microcontroller. Algorithms for detection of the peak of ECG signals are implemented using MCU to measure the heart rate, BPM, and IBI metrics. The json payload obtained from PubNub channels was published to freeboard.io subscription. Use of freeboard.io dashboard was used for visualization. It displays variation in IBI/HRV. Peak threshold determined is simple and accurate. Performance with moving average algorithm is better than with IBI. A similar heart monitoring system and database management system is proposed in [19]

9.3 VARIOUS HEALTHCARE MONITORING SENSORS AND EQUIPMENTS

9.3.1 Arduino Uno

The Arduino Uno is a microcontroller board designed by Arduino.cc that is based on the ATmega328 microprocessor and is regarded as the original Arduino board (Uno in Italian means "one").

Figure 9.1 shows the input/output pins, facility for USB connection, power supply reset, etc. The board has all the required support for the microcontroller.

9.3.2 Temperature Sensor

This sensor is comparable to a linear temperature sensor which gives a reading in Kelvin. Here there is no need to subtract a certain constant voltage from the output to get temperature in degree Celsius. Additional calibration is not required. Accuracies of ±¼°C and ±¾° cover the full range from –55°C to 150° (Figure 9.2).

9.3.3 Glucose Monitoring

IoT devices can help with these issues by offering continuous, automated glucose monitoring in patients. Glucose monitoring systems minimize the

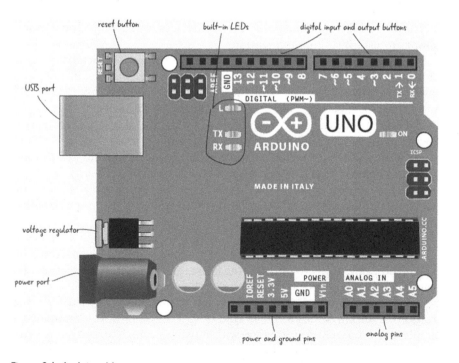

Figure 9.1 Arduino Uno.

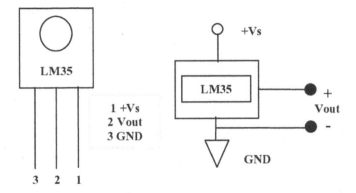

Figure 9.2 LM 35 temperature sensor.

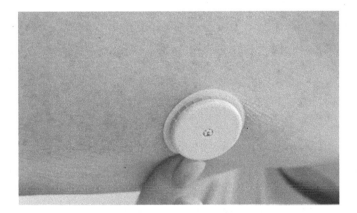

Figure 9.3 Glucose monitoring sensor.

need for manual record-keeping and can notify patients when their glucose levels are dangerously low (Figure 9.3).

Designing an IoT device for glucose monitoring that

a. is small enough to monitor constantly without causing patients any inconvenience is one of the challenges.
b. doesn't use so much power.

9.3.4 Heart-Rate Monitoring

These are plug-and-play devices used for sensing heart rate with Arduino. This is connected to Arduino or plugged to the finger or ear lobe using jumper wires. The front side having guts logo touches the skin. The LED shines through the small hole in it. It uses a 24-inch flat ribbon cable. The black wire resembles ground, the signal is given to purple wire and +3 V to +5 V supply is connected to red wire (Figure 9.4).

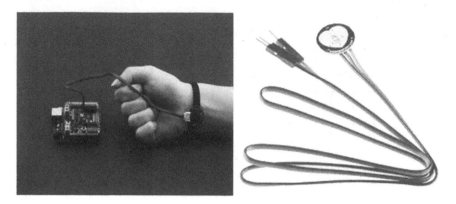

Figure 9.4 Heart monitoring pulse sensor.

9.3.5 Depression and Mood Monitoring

It is difficult to collect data regarding depression and mood monitoring. Healthcare employees are also not able to predict the mood changes in patients. IoT medical devices can help to predict the mood fluctuations in patients. In IoT medical devices, a smart band is available for the prediction of mood fluctuations (Figure 9.5).

9.3.6 Parkinson's Disease Monitoring

In order to treat Parkinson's patients effectively, healthcare providers must be able to assess how the intensity of their symptoms varies

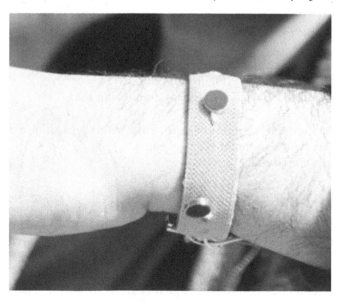

Figure 9.5 Smart band for emotions monitoring.

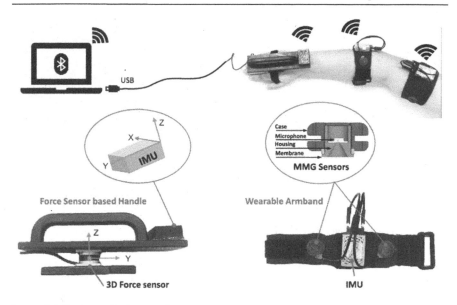

Figure 9.6 Parkinson's disease device.

throughout the day. IoT devices promise to make this procedure easier by continuously collecting data on Parkinson's symptoms. At the same time, the devices enable patients to live their lives at home rather than being confined to a hospital for lengthy periods of time for observation (Figure 9.6).

9.3.7 Ingestible Sensors

Internal parts of body are very sensitive and it is very difficult to collect data from internal body parts. And no one will allow inserting the camera inside the body. But with the help of an ingestible sensor, it is possible to collect data from the digestive and other systems in a far less intrusive method. This can give information about stomach's PII levels. In Figure 9.7 ingestible sensors are shown. This sensor is of very small size.

9.3.8 Robotic Surgery

Surgery is a difficult procedure that needs a great deal of expertise, surgical gear, medical technologies, and a trained surgeon. A surgeon must undertake a sophisticated process that is tough to control with hands during the surgical procedure. IoT robotic equipment can more accurately and effectively complete a surgical procedure. These gadgets are simple to insert and tiny enough to perform surgery in the patient's

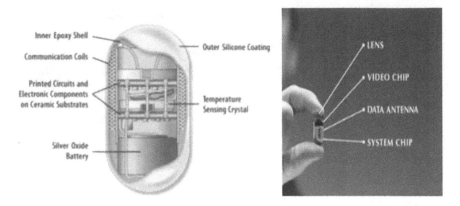

Figure 9.7 Ingestible sensors.

body with minimum interruption. IoT devices offer an unacceptable reality for healthcare cybersecurity firms in a linked healthcare network. Leaders in healthcare strive to comprehend what an organization entails, what it does, and how to manage and preserve everything (Figure 9.8).

9.3.9 Liquid Crystal Displays

LCDs are used to display various parameters. LCD shown in Figure 9.3 has 16 pins with 2 rows accommodating 16 characters each. These are often used in 4-bit mode or 8-bit mode. It has eight data lines and three control lines which will be used for control purposes (Figure 9.9).

9.3.10 ESP8266

The ESP8266 shown in Figure 9.10 is a low-cost WiFi microchip, with a full TCP/IP stack and microcontroller capability. It gives the application of WiFi connectivity.

9.3.11 Power Supply

The adapter used here provides 9 V and 1 Ampere DC output. In this work, we have used a power supply to give 12 V power to the hardware to function properly and provide the necessary power to all the components of the system.
Features:

- Incredibly low fault rates
- High reliability
- Regulated stable voltage
- Stable output
- Efficient and consumes less energy

Figure 9.8 Robotic surgery.

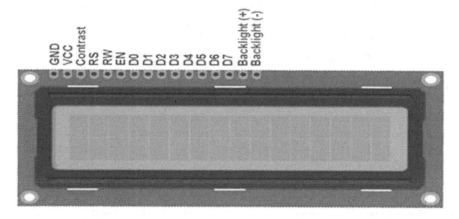

Figure 9.9 The LCD with its pin out.

Figure 9.10 ESP8266.

9.4 SOFTWARE SPECIFICATION

9.4.1 Arduino IDE

The Arduino integrated development environment (IDE) is written using functions from C and C++. It is a cross-platform application. Using third-party cores and development boards made by other vendors may also be used to upload programs to Arduino-compatible boards.

9.4.2 ThingSpeak (API)

Data collected from various sensors can be stored and retrieved using HTTP and MQTT protocol via LAN by use of ThingSpeak API. Various applications related to logging of sensors, tracking location, etc., can be created. ThingSpeak is used to analyze and visualize uploaded data using MATLAB. MQTT is the most widely used M2M/IoT protocol and has numerous applications and fetching real-time data.

9.5 CONCLUSION AND FUTURE WORK

The goal of the remote patient monitoring system was to make healthcare more inexpensive and accessible to the general public. The temperature and heart rate of the patient is taken. The readings are stored in the cloud (after being gathered via various sensors) and may be seen by the doctor or caretaker remotely from anywhere in the globe. This information can be utilized to analyse a patient's health and in medical investigations. The information can also be used in medical research investigations involving chronically unwell or elderly people. The expected result was achieved. The results of heartbeat, temperature sensing, and remote monitoring of the patient's vital signs, among other things, were as expected. In an emergency or critical situations, such as when a patient's temperature or heart rate deviates from desired values, or when a patient becomes ill and calls for help by pressing a button, alert signals to doctors and caregivers can help save their lives by prompting them to take the appropriate actions. It's also feasible to keep track of patient information on a regular basis.

Microcontrollers have advanced in speed, miniaturization, and power efficiency. They are simple to integrate into medical equipment. As a result, healthcare practitioners are turning to embedded systems to monitor their patients' health. People in most developing nations are currently utilizing their smartphones for such beneficial health applications, and they are also interested in new IoT gadgets that might help them in their daily lives. This suggests that individuals will begin to use IoT in the near future. People will benefit from a remote healthcare system, which will improve their quality of life. Better sensors may address additional vitals of a person in the future, making them more efficient and precise, and allowing them to be used to anticipate illnesses. In addition, an application can be created to make it easier to use the system more efficiently.

REFERENCES

[1] Gupta, N., Saeed, H., Jha, S., Chahande, M., & Pandey, S. (2017). Study and implementation of IOT based smart healthcare system. International Conference on Trends in Electronics and Informatics ICEI, Vol. 5, pp. 112, USA.

[2] Jaiswal, K., Sobhanayak, S., Mohanta, B. K., & Jena, D. (2017). IoT-cloud based framework for patient's data collection in smart healthcare system using Raspberry-pi. International Conference on Electrical and Computing Technologies and Applications (ICECTA).

[3] Reddy, N., Marks, A. M., Prabaharan, S. R. S., & Muthulakshmi, S. (2017). IoT augmented health monitoring system. IEEE, 12, 230–244.

[4] Sasidharan, P., Rajalakshmi, T., & Snekhalatha, U. (April 4–6, 2019). Wearable cardiorespiratory monitoring device for heart attack prediction. International Conference on Communication and Signal Processing, India.

[5] Saranya, E., & Maheswaran, T. PG scholar. (2019). IOT based disease prediction and diagnosis system for healthcare. IJEDR, 7(2), ISSN: 2321-9939.

[6] Sambhram, M. B. G. (Dec. 2016). Early prediction system of cardiovascular disease through IoT. International Journal of Combined Research & Development (IJCRD). eISSN: 2321-225X; pISSN: 2321-2241, 5(12).

[7] Banka, S., Madan, I., & Saranya, S. S. (2018). Smart healthcare monitoring using IoT. International Journal of Applied Engineering Research, ISSN 0973-4562, 13(15), 11984–11989. © Research India Publications. http://www.ripublication.com

[8] Ruman, Md. R., Barua, A., & Rahman, W. (Feb. 13–15, 2020). IoT based emergency health monitoring system. 2020 International Conference on Industry 4.0 Technology (I4Tech) Vishwakarma Institute of Technology, Pune, India.

[9] Jagannadha Swamy, T., & Murthy, T. N. (Jan. 23–25, 2019). eSmart: an IoT based intelligent health monitoring and management system for mankind. 2019 International Conference on Computer Communication and Informatics (ICCCI -2019), Coimbatore, India.

[10] Nookhao, S., Thananant, V., & Khunkhao, T. Development of IoT heartbeat and body temperature monitoring system for community health volunteer. 2020 Joint International Conference on Digital Arts, Media and Technology with ECTI Northern Section Conference on Electrical, Electronics, Computer and Telecommunications Engineering (ECTI DAMT & NCON), Vol. 12, pp. 1230–1241. USA.

[11] Tripathi, V., & Shakeel, F. (2017). Monitoring health care system using internet of things – an immaculate pairing. International Conference on Next Generation Computing and Information Systems (ICNGCIS), Vol. 112, pp. 130–141. USA.

[12] Kamal, N., & Ghosal, P. (2018). Three tier architecture for IoT driven health monitoring system using RaspberryPi. IEEE International Symposium on Smart Electronic Systems (iSES) (Formerly iNiS), Vol. 19, pp. 120–140. Singapore.

[13] Antonio, P. O. Rocio, C.M., Vicente, R., Carolina, B. , & Boris, B. (2017, October). Heat stroke detection system based in IoT. In 2017 IEEE Second Ecuador Technical Chapters Meeting (ETCM) (pp. 1–6). IEEE.

[14] Braam, K., Huangy, T. C., Cheny, C. H., Montgomery, E., Vo, S., & Beausoleily, R. Wristband vital: A wearable multi-sensor microsystemfor real-time assistance via low-power bluetooth link. 2020 International Conference on Industry 4.0 Technology (I4Tech)Vishwakarma Institute of Technology, Pune, India.

[15] Maria, A. R., Sever, P., & George, S. MIoT applications for wearable technologies used for health monitoring. ECAI 2018 – International Conference.

[16] Irmansyah, M., Madona, E., Nasution, A., & Putra, R. (2018). Low cost heart rate portable device for risk patients with IoT and warning system. International Conference on Applied Information Technology and Innovation (ICAITI).

[17] Kumar, G. V., Bharadwaja, A., & Nikhil Sai, N. (2017). Temperature and heart beat monitoring system using IOT. International Conference on Trends in Electronics and Informatics (ICEI). USA.

[18] Rajanna, R. R., Natarajan, S., & Vittal, P. R. (2017). An IoT Wi-Fi connected sensor for real time heart rate variability monitoring. 2018, IEEE Third International Conference on Circuits, Control, Communication and Computing. USA.

[19] Ren, Y., & Lyu, N. (April 28–30, 2016). A pulse measurement and data management system based on Arduino platform and android device. Proceedings of 2016 IEEE 13th International Conference on Networking, Sensing, and Control, Mexico City, Mexico.

Chapter 10

COVID-19 Patient Remote Health Monitoring System Using IoT

Jivan Parab, M. Lanjewar, C. Pinto, M. Sequeira, and G.M. Naik

CONTENTS

10.1 INTRODUCTION

The coronavirus disease (COVID-19), which started in Wuhan city in China, is a respiratory pandemic disease that created havoc in the entire world. This novel coronavirus leads to severe acute respiratory problems.

DOI: 10.1201/9781003355960-10

Li et al. defined COVID-19 as a suspected case of pneumonia which matches with below four criteria: (1) fever, with or without an increase in body temperature; (2) symptoms of pneumonia; (3) less or normal WBC count, and (4) symptoms remains even after treating with antimicrobial for three days [1]. As it respiratory disease, the main cause of fatality due to COVID-19 is the failure of hypoxic respiratory system [2,3]. COVID-19 has led to significant challenges in the medical fraternity as well as in social communities similar to the SARS-CoV outbreak in the year 2002–2003 and respiratory syndrome in the Middle East (MERS) in 2012 [4,5].

COVID-19 and several other viral diseases lead to physiological changes to some extent that can be monitored by wearable sensors. Some parameters generated from rhythmic heart beating such as heart rate (HR), heart rate variability (HRV), resting heart rate (RHR), and respiration rate (RR) could be used as an initial indication of COVID-19 infection and are being monitored by various wearable devices available in the market. The temperature of the body and SPO_2 are also monitored to check the fever and respiratory symptoms in COVID-19; however, these parameters are not regularly measured by the available wearables today. In patients who have tested positive for COVID-19, the infection leads to fluid-filled air sacs, which cause pneumonia. This translates to respiratory distress, where the patient experiences shortness of breath, chest pain, and rapid heartbeat [6]

HR varies with time which is nothing but a dynamic signal. The variation in HR signifies the current pathologic condition. HRV is defined as the variation in consecutive heartbeats period [7]. HRV measurement is a non-invasive indirect assessment of the control of autonomic nervous system on cardiac function and provides an idea about the balance in the activity between the sympathetic and parasympathetic nervous systems [8]. Viral illness increases body physiological stress which typically leads to an overall increase in HR. Spike in HR is also physiological response of the body during the defense against the infection [9]. A substantial decrease in HRV indicates not good recovery and indicates high level of physiological stress. Since there is a lack of clinical evidence on the predicted HRV during viral illness, there are also several self-reports and anecdotal proof which leads us to postulate that HRV trends can be used to predict the onset of illness [10]. Alteration in HRV is closely related to the presence of illness and pathologic conditions, and the degree of HRV alteration is an indicator of the severity of the current disease [11]. HR and HRV in a patient with COVID-19 have been studied. The patient with COVID-19 was associated with a decrease in HR and a paradoxical decline in HRV. An abrupt decline in HRV and a decrease in HR may signal the onset of COVID-19 before common symptoms such as dry cough or fever appear. In addition, HRV and HR measurements may help to evaluate the course of the disease [12]. The oxygen saturation (SpO_2) in a normal person should be above 94%. In the case of a COVID-19 patient, it drops below 94%. If SpO_2 falls below 94% an alert is generated and a message is sent to the monitoring team, comprised of respiratory physicians and nurse specialists [13]. Recently, researchers have proposed two innovative pathophysiology hypotheses (1) haemoglobin dysfunction and (2) tissue iron overload. There is a preliminary research about the heme metabolism of viral inhibition through surface glycoproteins by binding it to beta-chains of haemoglobin [14], this is then followed by one more publication about denaturation of haemoglobin [15]. This impact on haemoglobin will in turn lead to lesser SpO_2.

Recently, researchers have tried to analyze the respiratory rate to predict the risk of COVID-19 infection. They have created a model to find the changes in respiration rate in the case of COVID-19 patients. Along with all these parameters, the first and foremost symptoms are body temperature and fever [16]. The researchers have also studied the correlation between body temperature and mortality in COVID-19 patients [17].

Till date, there is no proven COVID-19-specific medical therapy, with basic monitoring of health parameters and supportive treatment which helps patients. Providing better patient care services round the clock is of utmost importance. Many researchers have designed systems which only monitor the utmost two parameters. Hence, we have aimed to develop a user-friendly low-cost COVID-19 patient remote health monitoring system which monitors RR, HR, SpO_2, HRV, total haemoglobin (tHb), and body temperature.

10.2 METHODOLOGY AND IMPLEMENTATION

The main objective of this paper is to design an affordable system for COVID-19 patients to remotely monitor important health parameters of patients such as body temperature, HR, HRV, SpO_2, tHb, and RR. These parameters along with GPS coordinates are uploaded every 15 minutes to ThingSpeak IoT server. Health care officers or district administration officers can access these vital health parameters by logging on to the ThingSpeak IoT server to take further action.

The block of the complete system design is shown in Figure 10.1. The main heart of the system is Espressif ESP32 Development Board, having a finger probe with three wavelengths of LEDs, a photodetector OPT101 which has an in-built Trans-impedance amplifier to acquire PhotoPlethysmography (PPG) to monitor HR, HRV, tHb, SpO_2, a piezoelectric sensor with signal conditioning to find RR, LM35 for sensing body temperature, and a GPS Module. The detailed implementation and acquisition are explained in the following sections.

10.2.1 Acquiring PPG Signal – HR, HRV, tHb, SpO_2

To measure HR, HRV, tHb, and SpO_2, we acquired PPG Signal using a finger probe which was designed with three LEDs having λ = 670, 810, and 950 nm on a single chip with a photodetector. The three LEDs were

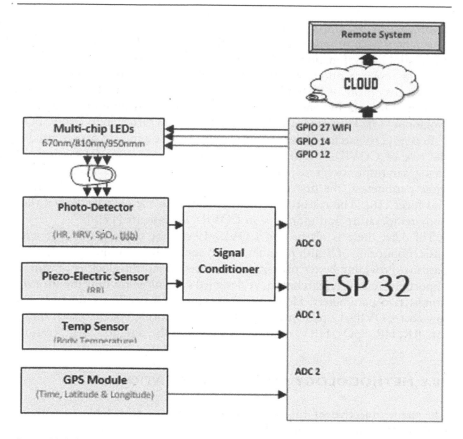

Figure 10.1 System block diagram.

sequentially turned on using ESP32 Development Board [18]. The PPG signal is a pulsating waveform which contains both AC and DC components. The pulsating AC component is because of the variation in the volume of blood during the systolic and diastolic states. The DC part is because of respiration, attenuation by epidermis, skin, tissue, and bone. During the systolic cycle, the diameter of the arteries increases thereby increasing the volume of blood. Due to this, the absorbance of light in tissues with arteries increases. During the diastolic cycle, the diameter of the arteries decreases thereby decreasing the volume of blood. Due to this, the absorbance of light in tissues with arteries decreases. This change in the volume of blood is a characteristic of the PPG waveform. These changes in the light intensity are detected by the OPT101 which is converted into varying PPG. The PPG signal is then filtered using an active bandpass filter (0.72–2.82 Hz) with a gain of 2. The amplified PPG signal is then digitized using a 12-bit ADC on ESP32. This information is then processed and the algorithm is executed to calculate HR, HRV, tHb, and SpO_2 in human blood.

10.2.1.1 Heart Rate from PPG Signal

Heart rate is the number of heartbeats in a minute. To calculate, HR from the sample PPG Signal shown in Figure 10.2, the time interval between adjacent peaks of the PPG signal is calculated [19].

$$PP_{av} = \frac{PP_1 + PP_2 + PP_3 + PP_4}{4} \tag{10.1}$$

where PP_{av} is the average of four PPG intervals between successive peaks of the PPG signal.

$$HeartRate = 60 * \frac{1}{PP_{av}} \tag{10.2}$$

10.2.1.2 Heart Rate Variability from PPG Signal

The changes in time intervals between adjacent heartbeats are nothing but HRV [20]. To calculate HRV from sample PPG Signal. Root mean square of successive differences (RMSSD) between normal heartbeats is found by first calculating time difference between each successive heartbeat. Then, all these values are squared and the average of the result is taken before taking the square root [20].

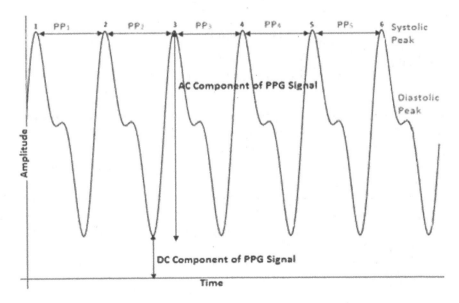

Figure 10.2 Sample PPG signal.

HRV (RMSSD)

$$= \sqrt{\frac{(PP_2 - PP_1)^2 + (PP_3 - PP_2)^2 + (PP_4 - PP_3)^2 + (PP_5 - PP_4)^2}{4}} \quad (10.3)$$

HRV value is calculated in terms of milliseconds.

10.2.1.3 Oxygen Saturation from PPG Signal

In order to calculate SpO$_2$, the AC and DC values of the pulsating RED (670 nm) and IR (950 nm) PPG are extracted and the ratio R is found. The ratio of ratios R is approximated in 10.4.

$$R = \frac{AC_{670}/DC_{670}}{AC_{950}/DC_{950}} \quad (10.4)$$

SpO$_2$ is computed based on the empirical calibration of the ratio of ratios for the specific device [21].

$$SpO_2\% = 115 - (15 \times R) \quad (10.5)$$

10.2.1.4 Total Haemoglobin from PPG Signal

Total haemoglobin is estimated by calculating the concentration of oxy-haemoglobin and reduced haemoglobin [22].

$$\begin{bmatrix} Hb \\ HbO_2 \end{bmatrix} = \begin{bmatrix} A_{670} \\ A_{810} \\ A_{950} \end{bmatrix} * \begin{bmatrix} \varepsilon Hb^{670} & \varepsilon HbO_2^{670} \\ \varepsilon Hb^{810} & \varepsilon HbO_2^{810} \\ \varepsilon Hb^{950} & \varepsilon HbO_2^{950} \end{bmatrix}^{-1} \quad (10.6)$$

where A_λ is the ratio of AC component of the PPG signal to DC component of the PPG signal for each wavelength when the finger is placed into the probe where λ = 670, 810, and 950 nm.

εHb^λ is the molar extinction coefficient for reduced haemoglobin for a particular wavelength.

$\varepsilon HbO_2^\lambda$ is the molar extinction coefficient for oxyhaemoglobin for a particular wavelength.

Hb and HbO$_2$ are the concentration of oxyhaemoglobin and reduced haemoglobin in moles per litre.

Total haemoglobin is obtained by adding oxyhaemoglobin and reduced haemoglobin.

$$tHb = Hb + HbO_2 \quad (10.7)$$

Figure 10.3 Placement of the piezo sensor.

10.2.2 Piezoelectric Sensors – RR

To measure the RR, we have designed a belt consisting of piezoelectric sensors wrapped around the chest. We have to position the piezo sensor on the right side of the chest as shown in Figure 10.3. The contraction and expansion of the chest during respiration will induce a mechanical vibration across the chest wall. These induced vibrations are then captured by the piezoelectric sensor which produces an electrical voltage [23]. This signal is then conditioned using an active low filter with a gain of 2, which is then given to the 12-bit ADC of ESP32. An algorithm is run on this digitized signal to find the number of peaks appear in one minute which is nothing but RR.

10.2.3 Sensor LM35 – Temperature

The LM35 sensor is a precision integrated-circuit temperature device whose output voltage increases linearly to the temperature change. The sensitivity of the LM35 is 10 mV per degree Celsius. LM35 is placed under the armpit to sense the body temperature. The output voltage of LM35 is then digitized using a 12-bit ADC of ESP32.

10.2.4 Global Positioning System (GPS) Module – Location

We have interfaced NEO GPS Module connected using UART interface to ESP32 Development Board. The GPS module fetches information from the satellite and then sends the information to the ESP32 in the NMEA (National Marine Electronics Association) format. This GPS information is sent along with all the health parameters to the ThingSpeak IoT server.

10.2.5 ThingSpeak for COVID-19 Patient Remote Monitoring

ThingSpeak is an IoT platform service that allows to analyze live data streams in the cloud. ThingSpeak collects and stores COVID-19 patient health parameters such as body temperature, HR, HRV, SpO_2, tHb, and RR. These data are uploaded to the IoT server every 15 minutes. In order to access and analyze these data, the doctor has to login to the ThingSpeak IoT server.

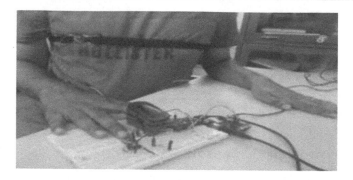

Figure 10.4 Setup of the implemented system.

The complete implemented system is shown in Figure 10.4.

10.3 SOFTWARE IMPLEMENTATION OF SYSTEM

Once all the sensors are placed on patients, the system is turned on for monitoring the health parameters. The first step is to configure three ADC channels to acquire sensor data of COVID-19 patients. The first channel i.e ADC Channel 0 is used to acquire PPG signals for three wavelengths of LEDs (670, 810, 950 nm). Next, the algorithm is run on PPG Signal to find peak and valley voltages from which AC and DC components of the PPG signal are determined. These AC and DC components are required to calculate HR, HRV, SpO_2, and tHb. Second, the ADC channel is configured to acquire the signal from the piezoelectric sensor to calculate RR. Third, the ADC channel reads the body temperature. Finally, all the parameters along with GPS coordinates are sent to the ThingSpeak IoT server after every 15 minutes. The flowchart depicting the detailed steps are shown in Figure 10.5. The entire coding is implemented using the MicroPython language.

10.4 RESULT AND DISCUSSION

The designed system is used to monitor the health parameters namely body temperature, RR, HR, HRV, SPO_2, and total haemoglobin which are essential in the case of COVID-19 patients. The designed system is validated with 14 subjects. The written consent of these subjects was taken for undergoing analysis during the measurement of the health parameters. First, the PPG signals were recorded using the designed finger probe, while the subjects were in the seated rest position. Each LED was turned on in a sequential pattern and kept ON for five seconds for acquiring the PPG signal with a one-second gap between the two PPG signals as shown in Figure 10.6.

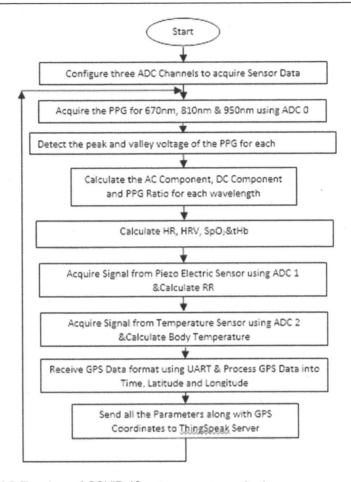

Figure 10.5 Flowchart of COVID-19 patient remote monitoring.

After recording PPG, next the signal from the piezoelectric sensor is recorded to find the respiration rate. The acquired signal is shown in Figure 10.7.

Table 10.1 gives the recorded data of the above health parameters of 14 subjects. The designed system was validated with reference readings recorded simultaneously using standard methods for these health parameters except for HRV and tHb. The validation of tHb is already published using three wavelengths [18], having the mean absolute error of 1.3778 gm/dL. We have not validated the HRV with a reference reading.

10.4.1 Bland–Altman Analysis

We use the coefficient of determination (r^2) and Bland–Altman plots to examine the agreement in the estimated and reference values of the above

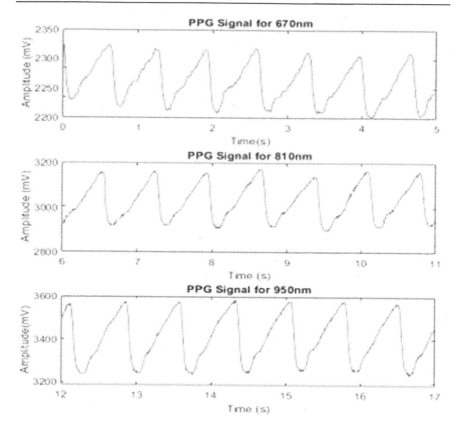

Figure 10.6 PPG signal.

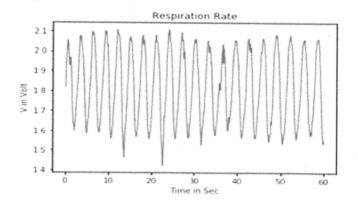

Figure 10.7 Piezoelectric signal from the chest.

Table 10.1 Measured health parameters

Subject No.	Gender	Body temperature (°F)		Respiration rate (Breath per min)		Heart rate (Beat per min)		Oxygen saturation (SpO₂)		Heart rate variability (HRV) (ms)	Estimated total haemoglobin (gm/dL)
		Estimated	Reference	Estimated	Reference	Estimated	Reference	Estimated	Reference		
Subject 1	M	98.1	98.0	17	17	79	80	97.12	97	20.19	15.80
Subject 2	M	97.2	97.4	18	18	66	65	97.19	97	33.53	15.44
Subject 3	M	97.0	96.9	20	20	63	64	96.50	97	52.83	14.25
Subject 4	F	97.2	97.8	17	17	64	64	97.64	98	20.55	13.24
Subject 5	F	96.8	96.8	18	18	79	80	97.17	97	53.09	11.80
Subject 6	M	96.8	97.0	21	20	99	100	98.70	99	24.69	13.86
Subject 7	F	96.8	96.6	20	20	87	88	96.70	97	46.72	14.10
Subject 8	M	98.1	98.4	22	21	78	78	97.80	98	93.45	12.23
Subject 9	F	97.0	97.2	18	18	73	73	97.23	98	33.76	10.20
Subject 10	M	98.3	98.4	19	19	72	72	97.37	98	62.71	12.30
Subject 11	F	98.1	98.0	20	20	77	77	98.90	99	31.25	11.32
Subject 12	M	96.8	97.2	21	21	90	91	98.56	99	81.51	14.87
Subject 13	M	96.3	96.4	20	20	62	63	98.40	98	35.84	13.56
Subject 14	M	98.0	98.2	19	18	89	89	97.66	98	39.02	15.53

health parameters monitored using the designed system. The Bland–Altman analysis graphically compares the estimated and reference values of a particular health parameter. It is a scatter plot of the difference between the estimated and the reference reading on the Y-axis and the corresponding mean of the estimated and reference measurements on the X-axis, [24]. Horizontal lines are drawn at the mean of the difference, and at the limits of agreement, which are defined ± 1.96 times the standard deviation (SD) from the mean of difference. Figure 10.8 shows the Bland–Altman graphs for four health parameters namely body temperature, RR, HR, and SpO_2.

It can be observed from Figure 10.8 that the mean of the difference between estimated and reference readings/standard deviation is –0.13/0.22, 0.21/0.43, –0.43/0.65, and –0.22/0.34 for body temperature, RR, HR, and SpO_2, respectively. From the Bland–Altman plots, it is clear that for the four health parameters of the subjects most of the readings lie within the limits of agreement. The mean of difference and SD for the above parameters clearly indicates that there is minimal spread in the values.

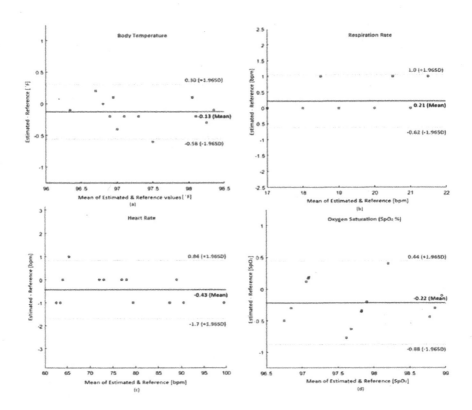

Figure 10.8 Bland–Altman plots.

10.4.2 Regression Analysis

The regression analyses are also done for these parameters and the results are plotted in Figure 10.9. All the parameters show good agreement. The regression analysis gives a coefficient of determination (r^2) of 0.90, 0.93, 1.00, and 0.81 for body temperature, RR, HR, and SpO_2, respectively as observed in Figure 10.9

10.4.3 Accuracy of the System

The accuracy of our system is calculated using 10.8. The accuracy of the health parameters is tabulated in Table 10.2.

$$Accuracy = \left(1 - \frac{\Sigma abs\,(Estimated - Reference)/Reference}{No.\ of\ Subjects} \right) \times 100$$

$$(10.8)$$

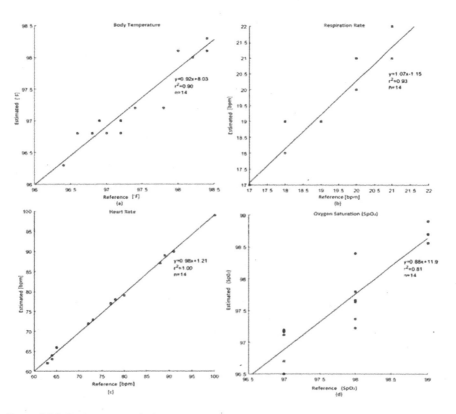

Figure 10.9 Regression analysis.

Table 10.2 Accuracy

S. No.	Parameter	Accuracy
1	Body temperature	99.80
2	Respiration rate	98.89
3	Heart rate	99.45
4	Oxygen saturation	99.78

10.4.4 Monitoring Parameters over IoT Server

The designed system measures the above health parameters using various sensors and the same is sent on the ThingSpeak IoT server every 15 minutes along with the GPS coordinates. Authorized concerned users can access the individual health parameters of the subjects to analyse the status. Figure 10.10 shows the measured health parameters of the subject uploaded on ThingSpeak Server every 15 minutes continuously.

10.5 CONCLUSION

We have designed a low-cost remote health monitoring system specially used for monitoring the vital parameters of COVID-19 patients. The health parameters of COVID-19 patients which we have measured are body temperature, SpO_2, HR, HRV, RR, and tHb. The most catchy feature of our system is that we are measuring the haemoglobin, HRV, and RR, which are normally not available in one system. In order to measure all these parameters, people have to use multiple devices. To test the efficacy of our system, we recorded these parameters of 14 subjects and all were estimated to have an accuracy of more than 98%. We have also validated and found that our designed system is complying with Bland–Altman analysis. The major challenge while designing is to acquire the PPG signal from which we can calculate the majority of parameters. All these parameters are then sent to the ThinkSpeak IoT server every 15 minutes and doctors can monitor the same just by logging into the server. The system is also embedded with a GPS feature, with which patients with high levels of deterioration in health conditions can be traced and brought to the nearby health centre for further treatment. The data of COVID-19 patients and their location also can be accessed by the district administration office on regular basis for their statistics and planning. The system can also be modified by putting some intelligence so that it can decide about the patient's health status and send an SMS or call the doctor.

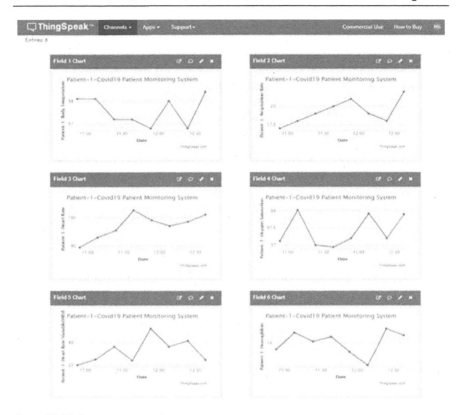

Figure 10.10 Remote monitoring.

Funding Source

There is no funding source.

Ethical approval

Written consent was obtained for all procedures performed in studies involving human participants.

Availability of data and material (data transparency)

Available but not provided here.

Code availability (software application or custom code)

Available but not provided here.

There is no conflict of interest.

REFERENCES

[1] Li, Q., Guan, X., Wu, P., Wang, X., Zhou, L., Tong, Y., et al. (2020). Early transmission dynamics in Wuhan, China, of novel coronavirus–infected pneumonia. *New England Journal of Medicine*, 382, 1199–1207. 10.1056/ NEJMoa2001316

[2] Belouzard, S., Chu, V. C., & Whittaker, G. R. (2009 April 7). Activation of the SARS coronavirus spike protein via sequential proteolytic cleavage at two distinct sites. *Proceedings of National Academy of Science USA*, 106(14), 5871–5876. 10.1073/pnas.0809524106 (accessed 19 April 2020).

[3] Yan, R., Zhang, Y., Li, Y., Xia, L., Guo, Y., & Zhou, Q. (2020). Structural basis for the recognition of SARS-CoV-2 by full-length human ACE2. *Science*, 367, 1444–1448. 10.1126/science.abb2762

[4] Paules, C. I., Marston, H. D., & Fauci, A. S. (2020). Coronavirus infections — more than just the common cold. *JAMA* (2020), 323, 707–708. 10.1001/ jama.2020.0757

[5] Fauci, A. S., Lane, H. C., & Redfield, R. R. (2020). Covid-19 — navigating the uncharted. *New England Journal of Medicine*, 382, 1268–1269. 10.1056/ NEJMe2002387

[6] American College of Cardiology. (2020). Troponin and BNP use in COVID-19. American College of Cardiology. Available online at: https:// www.acc.org/latest-in-cardiology/articles/2020/03/18/15/25/troponin-and-bnp-use-in-covid19 (accessed 20 April 2020).

[7] Saul, J. P. (1990). Beat-to-beat variations of heart rate reflect modulation of cardiac autonomic outflow. *News PhysiolSci*, 5, 32–33.

[8] Kleiger, R. E., Bigger, J. T., Bosner, M. S., Chung, M. K., Cook, J. R., et al. (1991). Stability over time of variables measuring heart rate variability in normal subjects. *American Journal of Cardiology*, 68(6), 626–630.

[9] Karjalainen, J., & Viitasalo, M. (1986). Fever and cardiac rhythm. *Archives of Internal Medicine*, 146, 1169–1171. 10.1001/archinte.146.6.1169

[10] Deusen, M. V. (2019). Heart rate variability: the ultimate guide to HRV. *WHOOP*. Available online at: https://www.whoop.com/thelocker/heart-rate-variability-hrv/ (accessed 12 April 2020).

[11] Hayano, J., Yamada, A., Mukai, S., Sakakibara, Y., Yamada, M., et al. (1991). Severity of coronary atherosclerosis correlates with the respiratory component of heart rate variability. *American Heart Journal*, 121(4 Pt 1), 1070–1079.

[12] Baumann, C., Buchhorn, R., & Willaschek, C. (2020). Heart rate variability in a patient with coronavirus disease 2019. *American Journal of Cardiology*. 2020050209. 10.20944/preprints202005.0209.v1.

[13] O'Carroll, O., MacCann, R., O'Reilly, A., Dunican, E. M., Feeney, E. R., Ryan, S., Cotter, A., Mallon, P. W., Keane, M. P., Butler, M. W., & McCarthy, C. (2020). Remote monitoring of oxygen saturation in individuals with COVID-19 Pneumonia. *European Respiratory Journal*. 10.11 83/13993003.01492-2020

[14] Wenzhong, L., & Hualan, L. (2020). COVID-19 disease: ORF8 and surface glycoprotein inhibit heme metabolism by binding to porphyrin. *ChemRxiv*. Preprint. 10.26434/chemrxiv.11938173.v3

[15] Wenzhong, L., & Hualan, L. (2020). COVID-19: attacks the 1-beta chain of hemoglobin and captures the porphyrin to inhibit human heme metabolism. *ChemRxiv*. Preprint. 10.26434/chemrxiv.11938173.v8

[16] Erenberk, U., Torun, E., Ozkaya, E., Uzuner, S., Demir, A. D., & Dundaroz, R. (2013). Skin temperature measurement using an infrared thermometer on patients who have been exposed to cold. *Pediatrics International, 55*, 767–770.

[17] Tharakan, S., Nomoto, K., Miyashita, S., & Ishikawa, K. (2020 June 5). Body temperature correlates with mortality in COVID-19 patients. *Critical Care, 24*(1), 298. 10.1186/s13054-020-03045-8

[18] Pinto, C., Parab, J., & Naik, G. (2020). Non-invasive hemoglobin measurement using embedded platform. *Sensing and Biosensing, 29*, 100373. 10.1016/j.sbsr.2020.100370

[19] McCraty, R., & Shaffer, F. (2015). Heart rate variability: new perspectives on physiological mechanisms, assessment of self-regulatory capacity, and health risk. *Global Advances on Health and Medicine, 4*(1), 46–61. 10.7453/gahmj.2014.073

[20] Shaffer, F., & Ginsberg, J. P. (2017). An overview of heart rate variability metrics and norms. *Frontiers of Public Health, 5*, 1–17. 10.3389/fpubh.2017.00258

[21] Shafique, M., & Kyriacou, P. A. (2012). Photoplethysmographic signals and blood oxygen saturation values during artificial hypothermia in healthy volunteers. *Physiological Measurements, 33*(12), 2065–2078. 10.1088/0967-3334/33/12/2065

[22] Sharma, C., Kumar, S., Bhargava, A., & Roy Chowdhury, S. (2013). Field programmable gate array based embedded system for non-invasive estimation of hemoglobin in blood using photoplethysmography. *International Journal on Smart Sensing and Intelligent Systems, 6*(3), 1267–1282. 10.21307/ijssis-2017-589

[23] Allataifeh, A., & Al Ahmad, M. (2020). Simultaneous piezoelectric non-invasive detection of multiple vital signs. *Scientific Reports, 10*(1), 1–13. 10.1038/s41598-019-57326-6

[24] Bland, J. M., & Altman, D. G. (1999). Measuring agreement in method comparison studies. *Statistical Methods in Medical Research, 8*(2), 135–160.

Chapter 11

Prediction of Alanine Using Neural Network Model

M. Sequeira, J.S. Parab, and G.M. Naik

CONTENTS

11.1 INTRODUCTION

Alanine is a vital amino acid, which is essential in the biosynthesis of proteins required by the human body [1]. Essential amino acids are those that the body is incapable of synthesizing on its own and must be obtained through food. A shortage of the above can lead to mental and physical health problems. Non-essential amino acids are those that the body can make on its own and hence are not needed to be supplied through food. Alanine is categorized as non-essential amino acid and all types of meats are a rich source of it. Other non-vegetarian sources for alanine are dairy items, eggs, various meat, seafood, gelatine as well lactalbumin [2]. Bran, corn, brown rice, brewer's yeast, whey, soy, beans, nuts, and legumes as well as whole grains are all good sources of alanine for vegetarians [2].

Alanine is ambivalent in nature; as such it can bond on both sides of the protein molecule. Another chemical property of alanine is that it is hydrophobic in nature [3]. It assists in the transport of toxins from the liver and conversion of the glucose into energy. Alanine plays a significant role in the maintenance of blood glucose levels in the body. Low tyrosine and phenylalanine levels, as well as high alanine levels, correlate to chronic fatigue syndrome and the Epstein–Barr virus. Alanine amino transferase

DOI: 10.1201/9781003355960-11

(ALT) is an important enzyme needed in the body. It is found in the cells of kidneys and liver, and in lower quantity in heart and muscle tissues of the body.

Alanine is essential in maintaining the glucose and nitrogen balance in the normally functioning human body by a chemical pathway known as alanine cycle. The alanine cycle is responsible to move the surplus amino acids in the tissue cells by converting them to pyruvate. Patients with liver diseases, diabetes, eating disorders, and low protein diets are advised by their physicians to take alanine supplements to avoid deficiency. Alanine aids in the absorption of vitamins B5 and B6, which are needed for overall health. There is a link between high alanine levels and high cholesterol, BMI, and blood pressure. Other factors that cause a rise in ALT are heart disease, bile duct obstruction, cirrhosis, and liver cancers [4].

Colorimetric [5,6], chromatographic [7], spectrophotometric [8,9], and chemiluminescence [10] techniques are commonly used to determine ALT amounts. However, such procedures necessitate the use of costly, heavy equipment as well as the involvement of qualified clinical personnel. Another technology exploited for ALT determination is paper-based analytical devices (PADs) [11–13]. Different approaches such as photolithography, handcrafting, cutting, and printing processes are employed in the fabrication of PADs [14–17].

However, all the above methods require withdrawal of blood for the proper implementation. In addition, the piercing site can be an occasion for an infection. This has generally caused apprehension in people's minds and led to less frequent monitoring of blood alanine. A non-invasive method circumvents all the above issues and is readily adaptable. Many researchers are working across the entire world to achieve a viable non-invasive method. In this research work, the authors explore the efficacy of the near-infrared region (NIR) for the prediction of alanine using the neural network method.

The light from 700 to 2,500 nm is termed NIR and the entire range is subdivided into three bands namely the combination band which extends from 2,000 to 2,500 nm, first overtone band which extends from 1,400 to 2,000 nm, and second or higher overtone band which extends from 750 to 1,400 nm. Many researchers have explored the combination region and brought out its utility for the prediction of various biomolecules such as glucose, urea, etc. [18–22]. This radiation in this region has a relatively deeper penetration in the skin than the NIR. The absorption peaks in the combination and overtone region are primarily attributed to the vibrational modes of C–H, O–H, and N–H bonds [23,24].

Various chemometrics techniques are essential to extract the alanine relevant signatures from the recorded dataset. Principal component analysis (PCA) and lately artificial neural network (ANN) are explored and documented in the literature for biomolecule prediction [25]. ANNs can model complex non-linear relations and are robust in nature and hence are used in quantitative and qualitative modelling [26,27]. ANN has consistently out-performed linear regression models for prediction [28,29]. Machine learning is explored for the application in colourimetric assay problems for detection. Machine learning regression as well as neural networks are deployed for efficient detection [30,31].

In this work, the authors have used the NIR absorption spectra of laboratory-prepared samples to explore the quantitative prediction of alanine in the sample. A machine learning model based on a neural network was developed to achieve the quantitative prediction of alanine. The method can be extended to estimate the alanine concentration in human blood.

11.2 METHODOLOGY

The analysis is based on the absorption of NIR light which can be determined using the Beer–Lambert law. The law states that there is a linear relationship between absorption due to the sample and the concentration along with sample thickness. ε is the constant of proportionality and is known as the molar extinction coefficient. The statement of the law is given as follows:

$$A = \varepsilon \cdot c \cdot l = \log\left(\frac{I}{I_0}\right) \tag{11.1}$$

The intensity of incident light is I_0, the intensity of remnant light after absorption is I, c is the analyte concentration present in the sample, ε is the molar extinction coefficient (L mol^{-1} cm^{-1}), and l is the thickness through which the light travels. If there are more than one chromophores in the absorbing tissue the Beer–Lambert law holds for each of the chromophores and the expression for it is given in the following equation:

$$A = (\varepsilon_1 \cdot c_1 + \varepsilon_2 \cdot c_2 + \varepsilon_3 \cdot c_3 \ldots \ldots \ldots + \varepsilon_n \cdot c_n) \cdot l \tag{11.2}$$

Here, n is termed as the number of chromophores present in the sample. If n chromospheres are present in the sample then c_n is the concentration of the nth chromophore and ε_n is the molar extinction coefficient for the nth chromophores.

11.3 EXPERIMENTAL SETUP

NIR spectra were recorded on spectrophotometer (JASCO V-770) over the wavelength range of 2,050–2,350 nm. It houses Czerny-Turner grating

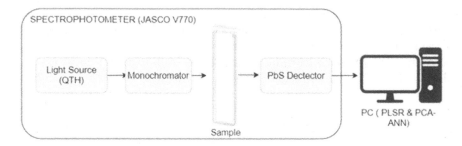

Figure 11.1 Block diagram of the experimental setup.

mount equipped with a QTH (Halogen) lamp (330–3200 nm) and a Peltier-cooled PbS detector for the NIR region. It operates in a dual beam mode. The spectrophotometer is operated by propriety spectra manager software (Figure 11.1).

11.4 SAMPLE PREPARATION

Laboratory samples were prepared containing alanine, glucose, ascorbate, urea, and lactate in an aqueous solution. These compounds are used as they resemble the blood matrix and were made to vary in concentrations relevant to the physiological range. The compounds were sourced from Sigma Aldrich Ltd. and were of analytical grade. The range of concentration variation for Alanine was 10–28 mg/dL, urea was 11–20 mg/dL, lactate was 12–22 mg/dL, glucose was 70–280 mg/dL, and ascorbate was 2–5 mg/dL. A total of 64 laboratory samples were prepared and NIR spectra were recorded of the same on the spectrophotometer. A quartz glass cuvette was chosen as the sample holder with a 1-mm path length.

11.5 DEVELOPMENT OF THE PCA–ANN MODEL

The PCA is used to analyze the covariance structures of the input absorption spectra [32,33]. By doing so, PCA gives a better way of visualization of the sources of variation in the original spectral data. Orthogonal vectors are extracted from the original data which characterize the variation and are referred to as principal components. Principal components have been interchangeably called loading vectors, spectral loadings, eigenvectors, etc.

Here, we have implemented PCA on the input spectral data and we used two principal components to model the input data, the total variance captured was 96.97%. The scores of these principal components are then fed to the ANN. This step reduces the dimensionality of the data as well as eliminates the noise and interference in the original spectral data. To implement PCA, we have used the command line function provided in MATLAB R2019a which is a part of the Statistics and Machine Learning Toolbox. Before the implementation of PCA, the data are centred. The PCA is implemented using the singular value decomposition (SVD) algorithm.

The machine learning is implemented using ANN. Levenberg–Marquardt's backpropagation algorithm was used to train the network. This is a part of the neural network fitting application which is bundled along with MATLAB R2019a. The hidden layers are constructed with neurons having a tan-sigmoid transfer function as the activation function. A linear activation function is used as the activation function for the neuron of the output layer. In addition to the input spectral data, the actual/reference reading of actual alanine present in the sample is given during the training phase. Once the network is trained we pass the prediction dataset to evaluate the prediction capability of the network.

11.6 PREDICTION USING PCA–ANN MODEL

To perform predictive analysis, machine learning using PCA–ANN was implemented. The networks modelled are shallow networks as there is only one hidden layer in the network along with one input and one output layer. The modelled network has two neurons in the input layer. The hidden layer has five neurons. The output layer has one neuron. The network is depicted in Figure 11.2. The inputs to the ANN are the scores of the loading extracted using the PCA in MATLAB and not the original absorption signatures. PCA helps to reduce the number of independent variables due to the data reduction features as well as eliminates the redundant interference and noise in the spectral absorption data. The recorded dataset was divided into training and validation set containing 57 and 7 samples, respectively. The former was used to train the model and the latter was used to test the prediction of the created model.

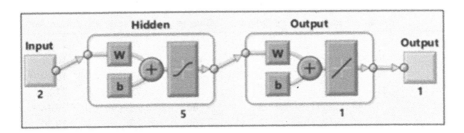

Figure 11.2 The structure of the trained ANN.

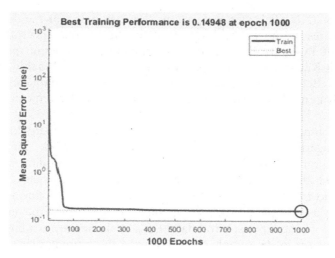

Figure 11.3 Training performance for PCA–ANN model.

The scores on these principal components for each training sample were passed to the input layer of the ANN. The training of the neural network was done on the 57 training samples and the mean squared error (MSE) is shown in Figure 11.3. MSE is 0.14948 at the 1,000th epoch. After the training, the prediction samples are transformed to the PCA space and sent to the trained ANN for alanine prediction. The prediction results are tabulated in Table 11.1.

The root mean square error for prediction of alanine is 0.44 and is calculated using the formula.

$$\text{Root Mean Squred Error} = \sqrt{\frac{1}{N}\sum_{i=1}^{N}(\hat{y}_i - y_i)^2} \qquad (11.3)$$

Table 11.1 Prediction of alanine using PCA–ANN

S. No.	Actual Concentration (mg/dL)					
	Urea	Lactate	Ascorbate	Glucose	Alanine	Predicted Alanine
1	20	12	5	70	28	27.87
2	20	12	5	100	10	10.11
3	11	22	2	200	10	10.69
4	11	12	5	280	28	27.25
5	20	22	5	280	28	27.74
6	20	12	2	200	28	28.30
7	11	12	5	100	10	10.31

11.7 VALIDATION

Validation is done by calculating the accuracy and by performing Bland–Altman analysis. It is a scatter plot where the differences between two measurements are plotted on the y-axis, and the corresponding mean of the two measurements is plotted on the x-axis. If the mean difference is closer, the measurement is said to be excellent. It is said that the method of measurement overestimates if the mean difference is above zero and tends to underestimate if it is below zero. Here a careful consideration is done for the spread of the limits of agreement, as wider limits of agreement signify a less precise method, whereas narrower limits indicate that the two methods agree well [34,35]. The bias is calculated by obtaining the mean of the difference between the actual and predicted readings and is calculated using equation (11.4) where b_a is the difference between the predicted value and actual value, and the value of n is the total number of subjects.

$$\text{Bias}(\bar{b}) = \frac{1}{n} \sum_{a=1}^{n} b_a \tag{11.4}$$

Standard deviation characterizes the variation in a set of values and can be obtained using the following equation:

$$\text{Standard deviation}(S_d) = \sqrt{\frac{1}{n-1} \sum_{a=1}^{n} (b_a - \bar{b})^2} \tag{11.5}$$

Another quantity called as the limits of agreement includes 95% of the differences between the two quantities namely the actual and predicted reading. The quantity can be obtained using equation (11.6). The limits are set as $\pm$ 1.96 times the standard deviation (S_D) from the mean of difference.

$$\text{limits of agreement} = \text{bias} \pm 1.96\, S_d \tag{11.6}$$

Bland–Altman analysis for prediction using PCA–ANN is plotted in Figure 11.4. The observed bias is shown to be 0.04 mg/dL and the limits of agreement are from -0.88 to 0.95 mg/dL. In PCA–ANN modelling, Bland–Altman plot and analysis showed very low bias with narrower limits of agreement which indicates the system is precise.

Accuracy is an important parameter required to characterize the method of prediction and can be obtained using equation (11.7) and is found to be 97.68%.

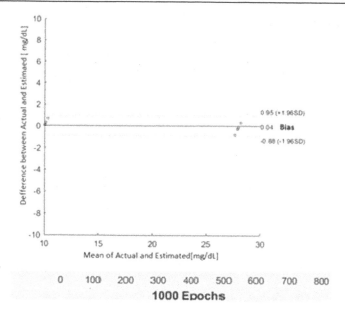

Figure 11.4 Bland–Altman Plot for PCA–ANN model.

$$\text{Accuracy} = \left(1 - \frac{\sum abs(\text{Predicted} - \text{Actual})/\text{Actual}}{\text{No. of subjects}}\right) * 100 \qquad (11.7)$$

11.8 CONCLUSION

The alarming rate of liver diseases all over the world has called for a reliable and accurate non-invasive method for the prediction of alanine in the complex blood matrix. The authors performed a systematic investigation to estimate alanine using machine learning using PCA–ANN. Authors have successfully shown that alanine prediction can be done using absorption signatures in the range of 2,050–2,350 nm. The PCA–ANN gave an RMSE of prediction of 0.43 and an accuracy of prediction 97.68%. The Bland–Altman plots clearly show that the prediction of alanine using PCA–ANN has a very low bias and narrow limits of agreement making the method precise. Hence, PCA–ANN is an excellent method for non-invasive alanine prediction.

REFERENCES

[1] Alberts, B., Johnson, A., Lewis, J., Raff, M., Roberts, K., & Walter, P. (2002). *Molecular biology of the cell*. 4th ed. USA: Garland Science.

[2] Nazareth, I. A. P., Vernekar, S. R., Gad, R. S., & Naik, G. M. (2016). Response of cholesterol in the presence of alanine and other blood constituents. *International Journal of Electronics and Communication Engineering & Technology*, 7(1), 101–106.

[3] Nilsson, I., Johnson, A. E., & von Heijne, G. (Aug. 2003). How hydrophobic is alanine? *Journal of Biological Chemistry*, 278(32), 29389–29393. 10.1074/jbc.M212310200

[4] The amino acid alanine – benefits, information on supplements, articles, links, news, advice. http://www.vitaminstuff.com/amino-acid-alanine.html (accessed 14 March 2022).

[5] Babson, A. L., Shapiro, P. O., Williams, P. A. R., & Phillips, G. E. (Mar. 1962). The use of a diazonium salt for the determination of glutamic-oxalacetic transaminase in serum. *Clinica Chimica Acta*, 7(2), 199–205. 10.1016/0009-8981(62)90010-4

[6] Lippi, U., & Guidi, G. (June 1970). A new colorimetric ultramicromethod for serum glutamicoxalacetic and glutamic-pyruvic transaminase determination. *ClinicaChimicaActa*, 28(3), 431–437. 10.1016/0009-8981(70)90069-0

[7] Canepari, S., Carunchio, V., Girelli, A. M., & Messina, A. (June 1994). Determination of aspartate aminotransferase activity by high-performance liquid chromatography. *Journal of Chromatography B Biomedical Applications*, 656(1), 191–195. 10.1016/0378-4347(94)00069-7

[8] Karmen, A. (Jan. 1955). A note on the spectrometric assay of glutamic-oxalacetic transaminase in human blood serum. *Journal of Clinical Investigation*, 34(1), 131–133.

[9] Itoh, H., & Srere, P. A. (June 1970). A new assay for glutamate-oxaloacetate transaminase. *Anal of Biochemistry*, 35(2), 405–410. 10.1016/0003-2697(70)90202-2

[10] Janasck, D., & Spohn, U. (Feb. 1999). Chemiluminometric flow injection analysis procedures for the enzymatic determination of L-alanine, α-ketoglutarate and L-glutamate. *Biosensors and Bioelectronics*, 14(2), 123–129. 10.1016/S0956-5663(98)00115-8

[11] Martinez, A. W., Phillips, S. T., Butte, M. J., & Whitesides, G. M. (2007). Patterned paper as a platform for inexpensive, low-volume, portable bioassays. *AngewChemInt Ed Engl*, 46(8), 1318–1320. 10.1002/anie.200603817

[12] Bruzewicz, D. A., Reches, M., & Whitesides, G. M. (May 2008). Low-cost printing of poly(dimethylsiloxane) barriers to define microchannels in paper. *Analytical Chemistry*, 80(9), 3387–3392. 10.1021/ac702605a

[13] Pollock, N. R., et al. (Sep. 2012). A paper-based multiplexed transaminase test for low-cost, point-of-care liver function testing. *Science Translational Medicine*, 4(152), 152ra129. 10.1126/scitranslmed.3003981

[14] Martinez, A. W., Phillips, S. T., Wiley, B. J., Gupta, M., & Whitesides, G. M. (Dec. 2008). FLASH: a rapid method for prototyping paper-based micro-fluidic devices. *Lab Chip*, 8(12), 2146–2150. 10.1039/b811135a

[15] Lu, Y., Shi, W., Jiang, L., Qin, J., & Lin, B. (May 2009). Rapid prototyping of paper-based microfluidics with wax for low-cost, portable bioassay. *Electrophoresis*, 30(9), 1497–1500. 10.1002/elps.200800563

[16] Nie, J., Liang, Y., Zhang, Y., Le, S., Li, D., & Zhang, S. (2013). One-step patterning of hollow microstructures in paper by laser cutting to create

microfluidic analytical devices. *Analyst*, *138*(2), 671–676. 10.1039/C2AN3 6219H

[17] Wang, H.-L., Chu, C.-H., Tsai, S.-J., & Yang, R.-J. (Jan. 2016). Aspartate aminotransferase and alanine aminotransferase detection on paper-based analytical devices with inkjet printer-sprayed reagents. *Micromachines (Basel)*. *7*(1), E9. 10.3390/mi7010009

[18] Kang, N., Kasemsumran, S., Woo, Y.-A., Kim, H.-J., & Ozaki, Y. (May 2006). Optimization of informative spectral regions for the quantification of cholesterol, glucose and urea in control serum solutions using searching combination moving window partial least squares regression method with near infrared spectroscopy. *Chemometrics and Intelligent Laboratory Systems*, *82*(1), 90–96. 10.1016/j.chemolab.2005. 08.015

[19] Hazen, K. H., Arnold, M. A., & Small, G. W. (Oct. 1998). Measurement of glucose and other analytes in undiluted human serum with near-infrared transmission spectroscopy. *Analytica Chimica Acta*, *371*(2), 255–267. 10.1016/S0003-2670(98)00318-3

[20] Parab, J., Sequeira, M., Lanjewar, M., Pinto, C., & Naik, G. (2021). Backpropagation neural network-based machine learning model for prediction of blood urea and glucose in CKD patients. *IEEE Journal of Translational Engineering in Health and Medicine*, *9*, 4900608. 10.1109/ JTEHM.2021.3079714

[21] Chen, J., Arnold, M. A., & Small, G. W. (Sep. 2004). Comparison of combination and first overtone spectral regions for near-infrared calibration models for glucose and other biomolecules in aqueous solutions. *Analytical Chemistry*, *76*(18), 5405–5413. 10.1021/ac0498056

[22] He, Y., Li, X., & Deng, X. (April 2007). Discrimination of varieties of tea using near infrared spectroscopy by principal component analysis and BP model. *Journal of Food Engineering*, *79*(4), 1238–1242.

[23] Burns, D. A., & Ciurczak, E. W. (2008). *Handbook of near-infrared analysis*. Boca Raton, FL: CRC Press.

[24] Siesler, H. W. (ed.) (2002). *Near-infrared spectroscopy: principles, instruments, applications*. Weinheim: Wiley-VCH.

[25] Tenhunen, J., Kopola, H., & Myllylä, R. (Oct. 1998). Non-invasive glucose measurement based on selective near infrared absorption; requirements on instrumentation and spectral range. *Measurement*, *24*(3), 173–177. 10.1016/ S0263-2241(98)00054-2

[26] Yu, S., Zhu, K., & Diao, F. (Jan. 2008). A dynamic all parameters adaptive BP neural networks model and its application on oil reservoir prediction. *Applied Mathematics and Computation*, *195*(1), 66–75.

[27] Haykin, S. *Neural Networks and Learning Machines*, vol. 3. USA: Upper Saddle.

[28] Yeh, S.-J., Hanna, C. F., & Khalil, O. S. (June 2003). Monitoring blood glucose changes in cutaneous tissue by temperature-modulated localized reflectance measurements. *Clinical Chemistry*, *49*(6Pt 1), 924–934. 10.1373/ 49.6.924

[29] Ming, C. Z., & Raveendran, P. (2009). Comparison analysis between PLS and NN in noninvasive blood glucose concentration prediction. *2009 International Conference for Technical Postgraduates* (pp. 1–4).

[30] Kim, H., Awofeso, O., Choi, S., Jung, Y., & Bae, E. (2017). Colorimetric analysis of saliva-alcohol test strips by smartphone-based instruments using machine-learning algorithms. *Applied Optics*, *56*, 84. 10.1364/ao.56.000084

[31] Mutlu, A. Y., Kiliç, V., Ozdemir, G. K., Bayram, A., Horzum, N., & Solmaz, M. E. (2017). Smartphone-based colorimetric detection: via machine learning. *Analyst*, *142*, 2434–2441.

[32] Wold, S., Esbensen, K., & Geladi, P. (Aug. 1987). Principal component analysis. *Chemometrics and Intelligent Laboratory Systems*, *2*(1), 37–52. 10.1016/0169-7439(87)80084-9

[33] Nomikos, P., & MacGregor, J. F. (1994). Monitoring batch processes using multiway principal component analysis. *AIChE Journal*, *40*(8), 1361–1375. 10.1002/aic.690400809

[34] Giavarina, D. (2015). Understanding Bland Altman analysis. *Biochemia Medica*, *25*(2), 141–151. 10.11613/BM.2015.015

[35] Bland, J. M., & Altman, D. G. (Feb. 1986). Statistical methods for assessing agreement between two methods of clinical measurement. *The Lancet*, *327*(8476), 307–310. 10.1016/S0140-6736(86)90837-8

Chapter 12

Hybrid Algorithm for Detection of COVID-19 Using CNN and SVM

Pranjali G Gulhane and Anuradha Thakare

CONTENTS

12.1 INTRODUCTION

Novel coronavirus spread over the globe leads to the necessity to detect it early and provide proper treatment for it. The pandemic waves are coming regularly and the detection methods are still the same as the first wave. The patient is first sent for an RT-PCR test. The result of this test takes more than 24 hrs depending upon the lab availability. This congenital test helps to detect only the presence of COVID-19. Then the patient is treated if positive, and the vital time of treatment is lost. Therefore different solutions to the detection of infection are suggested. Here we develop the software for COVID-19 detection using AI; specially designed for front-line use to help doctors to detect and monitor the disease efficiently and effectively. Patients with confirmed COVID-19 pneumonia have typical imaging features that can be helpful in the early screening of highly suspected cases and in the evaluation of the severity and extent of the disease. Most patients with COVID-19 pneumonia have ground-glass opacities or mixed ground-glass opacities and consolidation and vascular enlargement in the lesion. Lesions are more likely to have

DOI: 10.1201/9781003355960-12

peripheral distribution and bilateral involvement and be lower X-Rays predominant and multifocal. CT involvement score can help in the evaluation of the severity and extent of the disease [1].

Some surveys analyzed that the sensitivity of RT-PCR testing at various tissue sites and bronchoalveolar lavage fluid specimens demonstrated the highest positive rates at 93% ($n = 14$) [2]. This was followed by sputum at 72% ($n = 75$), nasal swabs at 63% ($n = 5$), fibro bronchoscope brush biopsy at 46% (6/13), pharyngeal swabs at 32% ($n = 126$), faeces at 29% ($n = 44$), and blood at 1% ($n = 3$). The authors of that study pointed out that testing of specimens from multiple sites may improve the sensitivity and reduce false-negative test results. The letter examined 1,070 specimens that were collected from 205 hospitalized patients with confirmed COVID-19 in China.

In another study published in radiology, investigators found chest CT achieved higher sensitivity for the diagnosis of COVID-19 as compared with initial RT-PCR from pharyngeal swab samples [3]. This retrospective study analyzed 1,014 hospitalized patients with suspected COVID-19 in Wuhan, China, with patients undergoing both serial RT-PCR testing and chest CT. Using RT-PCR results as reference standard, the sensitivity, specificity, and accuracy of chest CT in diagnosing COVID-19 were 97% ($n = 580$), 25% ($n = 105$), and 68% ($n = 685$), respectively.

These studies show the importance of X-ray images for the detection of the severity of infection. Instead of RT-PCR testing, which has low correctness, the exact infection level can be detected by X-ray images. The volume of patients admitted can be controlled by knowing the exact state of infection. This will give another advantage in social financial spending during the pandemic. Patients with severe infections can be treated with costly medicine and all others can be cured by regular medicine at home. Our solution is fast and takes into consideration of possible mistakes during image processing. The X-ray images collected are not uniform and need to be corrected by augmentation. Which takes care of the deformation of images collected. The X-ray images have a lot of noise which is necessary to remove. So extensive proper pre-processing of X-ray images will increase the accuracy of automated detection process. The uniform images are then processed for ground-glass opacities. Patients with a specified threshold will be treated in house or with mixed ground-glass opacities. The images are classified using the support vector machine (SVM) technique for detection of the level of infection. Only patients with a specified threshold will be treated in the hospital to control the infection.

12.2 LITERATURE SURVEY

Garlapati et al. in their paper [1] propose an automatic detection system based on lung X-ray images, as radiography modalities are a promising way of faster diagnosis. The black patches in images give the information about the level of infection in the lungs. In this work, a machine learning model is built, considering X-ray images taken from publicly available datasets of 2,000 images. Proper segmentation of patches in the images is done after the removal of noise from the images. The Sobel filter is used for proper edge detection of patches. The image augmentation images are also added in training in order to make the model more reliable. The relevant features are extracted from the images for building of the model. The X-ray images are prone to noise and spatial aliasing which leads the boundary to be indistinguishable, so proper boundary detection is necessary for the segmentation of image. For the classification of images in covid and non-covid classes, SVM kernel is used. This SVM kernel gives high accuracy and precision. Almost 99% accuracy is observed in given data images.

Hany et al. in their research [2] proposed a new method based on ensembles with two-stage classification for the detection of positive and negative cases. This will help in the early detection of COVID-19 and mild cases can be detected for proper treatment. The early detection of covid cases will help patients to save financial losses due to unwanted treatment in mild cases. Their study is based on regular blood tests collected at labs. The training and testing are done on almost 200 cases with known results. The performance of existing classification methods such as Gradient Boosting and random forest can be improved by applying a hybrid approach. The symptoms are used as parameters for classification. After applying gradient boosting classifier, similarity distance is checked. This ensemble similarity index gives more accuracy in the classification of COVID-19 cases as positive and negative cases. Supervised learning as well as clustering techniques are used. K-mode clustering is used to form two classes as positive and negative. On each cluster, classification models such as Random Forest and gradient boosting are applied. The two stages work in a sequential fashion so that accuracy will improve. Ensemble voting approach is applied to find the maximum accuracy model out of model created by cross-validation. The authors found that random forest gives 87% accuracy, which is further increased to 90% by the ensemble approach.

Oliveira and Bastos-Filho [4] proposed the method to apply machine learning methods to classify the traditional laboratory test parameters to detect positive cases. The laboratory parameters are considered for classification of patients so that hospitalization can be decided earlier. The time taken by traditional analysis of clinical test takes a lot of time and disease take this time to spread within a patient and the community around the patient. The disease prediction model is built around methods such as Random Forest, Multi-Layer Perceptron, and SVMs Regression. This will reduce the time to analyse the clinical parameters. They have developed a two-stage strategy based on the correlation between clinical parameters and COVID-19 detection. In the first stage, it detects the COVID-19 presence by machine leering algorithm and the second stage supports the decision on hospitalization. Thereby controlling the treatment time and spread of the virus.

Koushik and Bhattacharjee [3] adopted the approach in which also they use clinical tests for the detection of COVID-19 using machine learning techniques. In this paper, blood test parameters such as CBC, CRP, D-Dimer, S-ferritin, ALT, and LDH are considered for analysis. The early detection of COVID-19 may control the death rate and treatment may be more effective. So authors have used blood parameters to detect COVID-19 instead of PCR parameters. PCR parameters are difficult to collect and the time required for analysis is more. Meantime disease may become severe and the patient may need hospitalization. The blood test helps to discriminate between COVID-19 and other pneumonia. The economic advantages of such methodology are also discussed by authors. Authors used different approach of Machine Learning technique with different classifiers such as Random Forest (RF), SVM, and Naive Bayes. The authors applied all three tests on the dataset collected by them and found SVM classifier performance is satisfactory as compared to other classifiers. Almost 88% of accurate results are found detecting correct results.

Rahm and Kurniawan et al. in [5] have taken the issue of hospitalization during the COVID-19 pandemic. As capacity of infrastructure was getting full due to huge number of patients, it is necessary to prioritize the admission of patients to the hospital based on their infection. In this paper, a novel decision-making system is proposed to determine the patient's priority in admission to the hospital. The same technique is also useful to detect covid and non-covid patients based on X-ray images of chest. The decision-making system based on AI algorithm can support the hospital management to take automated decision for the admission of patients. For automation instead of RT-PCR clinical test, X-ray machines are used to provide the images of lungs. Normal laboratory result of RT-PCR takes time of more than 24 hours to declare the result of positive and negative case. VGG-16 network is used to classify the images in covid and non-covid patients. The X-ray images collected are used to train the machine learning algorithms used. From the images collected by X-ray, machine learning algorithm detects the severity of infection. This level of severity will help the decision-making system to take automated process to select the patient for admission in hospital. VGG-16 is one of the types of convolution neural network, which is primarily used for classification. The classified results are passed to different criteria and sub-criteria to support the decision-making system. The analytic hierarchy process (AHP) model is used by decision support system to take decisions. The consistency index will make a final decision of admission by passing through criteria and sub-criteria already defined. The recall and precision were found to be 96% for correct decision in their experiments on different datasets such as squeezenet, densenet, inception, and resnet. Miah and Yousuf presented a study [6] on lung cancer based on CT images. Their study used a neural network–based algorithm for the detection of lung infection for cancer. COVID-19 also creates infection in the lungs. This study is helpful for applying similar technique to know COVID-19 cases from lung images. Banalization is applied to images and then threshold is applied to detect the cancer and non-cancer cases. Steps of pre-processing, banalization, and thresholding are done before the neural network classifier. Feature extraction is done by segmentation of both left and right lungs from original images. Almost 96% detection rate is achieved by classifier. The drawback of this technique is the processing time taken at neural network classifier.

Reshi et al. [7] proposed the method of deep learning for X-ray image classification for covid and non-covid patients. For better training of convolutional neural network (CNN) from a small set of images, they have employed the technique of image augmentation. The data augmentation technique is an efficient way to increase the data size and quality of database. Dataset balancing and expert opinion are included in data pre-processing to increase the quality of images. X-ray images are not always clear and it is difficult to train the CNN on such low-quality images. Image augmentation overcomes this problem and gives efficient training necessary for CNN [8]. To overcome the drawback of every technique, authors have proposed different techniques for each stage of pre-processing, image augmentation, and classification. The CNN training is incrementally performed with different datasets. This helps in attaining the maximum accuracy and performance. In incremental training depending upon the performance matrix, each layer is added after every iteration. This is continued until stable performance is achieved. Limited size of primary dataset and class distribution imbalanced issues with the primary dataset affected the performance of the CNN models [9]. Image augmentation overcomes this issue to achieve 98% performance in classification. Loddo et al. [10] in their research of deep neural networks for COVID-19 detection have compared public datasets available. Two out of the available public dataset are taken for experiment. Ten popular convolution neural networks are taken for study and their performance is compared [11]. The authors found that VGG19 is best suitable for the application of covid classification. The cross-dataset experiment showed that deep learning networks have only 75% performance in real-time application for covid patient classification [12,13]. Their observation shows that it has poor performance for heterogeneous image sources.

12.3 PROPOSED METHOD AND ALGORITHM (FIGURE 12.1)

12.3.1 System Architecture

Our system's purpose is to design a system which can help hospitals to identify covid patients in a quick manner and also to know the intensity of infection so that decision of admission can be taken. For that, X-ray images of the chest are taken as input to the covid detection system. Based

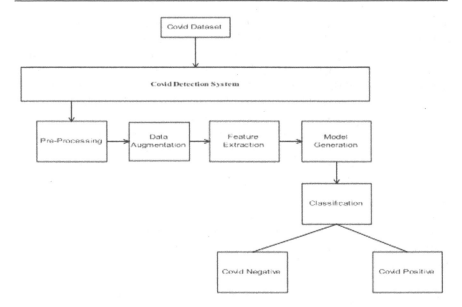

Figure 12.1 Block diagram of a hybrid approach.

on the images, the system will detect positive patients along with intense infections. This is will reduce two-stage testing of patients. In the normal flow of diagnosis, the patient is first passed to RT-PCR test, the result of which takes more than 24 hrs. It doesn't show the level of infection, hence again X-ray images are required. The analysis of these images gives the level of infection and then the patient is classified as severe or low in-fection. Our system's aim is to reduce this delay and reduce the crowd of admission of all patients to the hospital.

For this purpose, we are using X-ray images as input to our system. The analysis of X-ray images serves two purposes: detection of positive cases and detection of the level of infection. Only patients with a higher level of infection are isolated and treated in hospital. When X-ray images are passed through pre-processing stage, their aspect ratio and intensity are normalized. Images are not uniform therefore augmentation stage is in-cluded. This augmentation stage brings all images in normalized form, so that their features can be extracted. The black patches in the images are highlighted by means of the Sobel edge detection algorithm. Such an image is passed through CNN stage so that features can be extracted in numerical values. In the last stage, these images are passed to SVM classifier to detect positive and negative cases. This method is reliable and fast as only X-ray snaps of patient's chest is sufficient. And it gives the level of infection also, which leads to immediate hospitalization of serious patients. These fast detection and isolation techniques are helpful to re-duce the death rate in COVID-19.

12.3.2 Dataset

For experimental work, we have used image dataset from Github [14]. The images are selected by supervised medical help. Out of 700 images available in the dataset, only 135 images are selected as confirmed covid cases. The remaining 45 images are selected as non-covid cases. The CXR images contain covid and non-covid samples. As the dataset volume for exact COVID-19 positive cases is less, image augmentation is done. This help during development in two way. It increases the count of images and increases the quality of images. This dataset is used for training and testing our algorithm and proposed integrated system.

12.3.3 Pre-processing (Figure 12.2)

In the pre-processing stage, it is necessary to normalize the images. The images from the dataset are converted into greyscale. This will only illuminate information of images. The size of images also gets reduced which helps in speeding up of pre-processing stage and next image processing stages. In the normalization stage, all images are converted into 150 × 140 pixels irrespective of their initial size. By applying filters such as the median, noise reduction is done. More noise images give wrong training to machine learning algorithms.

12.3.4 Data Augmentation

To increase the data instances from available data, the data augmentation technique is used. [8]. The increased instances will help to train the model efficiently. For our image dataset, data augmentation can be performed by basic image operations. These operations are flipping, rotating, cropping, or padding. The new dataset formed by these operations helps to increase the training samples and thereby CNN module becomes effective. To overcome the limitation of dataset size for training, data augmentation is used in our study to train CNN. This is also used to provide more features

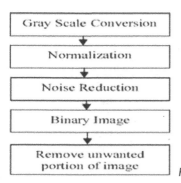

Figure 12.2 Pre-processing of images.

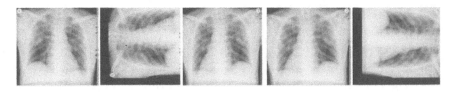

Figure 12.3 Flip and rotate operation on CXR images.

for the model. For our data augmentation, we have used only two techniques flip and rotation. In the first operation, X-ray images are flipped to get one more set of images. In the second stage, the original image is rotated by 90° to get the third dataset. In the third stage, the original image is rotated by 180° to get the fourth dataset. In the last stage, the original image is rotated by 270° to get the fifth dataset. Data augmentation overcomes the image variance problem by taking images from different angles. This also helps during the classification stage and class differences get magnified (Figure 12.3).

Table 12.1 lists the operations performed on the image during augmentation. The count of images resulted from the different operations. It increases the size of the image database. As well as the quality of images gets improved for the training purposes of CNN.

12.3.5 Convolution Neural Network (Figure 12.4)

A convolution neural network (CNN) works as the human brain [15]. CNN is used in image processing because their processing can be equal to the human viewing systems. Therefore, CNN is used in image processing, image analysis, and image classification. CNN consists of a convolution layer, max pooling, and non-linear activation layer. ReLU is used as an activation function as it increases the non-linearity in the input layer because the image is non-linear in nature. CNN implemented with ReLU is easier and faster. As data augmentation is performed, it enhances the quality and size of the dataset. Such a dataset is necessary for better model to envelop in deep learning.

Table 12.1 Training data of CNN

Operation	Image count
Original	180
Original flipped	180
Original with a 90° rotation	180
Original with 180° rotation	180
Original with 270° rotation	180
Total dataset after augmentation	900

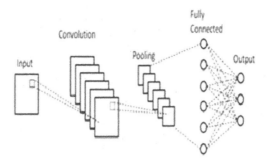

Figure 12.4 CNN block diagram.

12.3.6 SVM Classifier (Figure 12.5)

For classification, we have used SVM. SVM takes input from the CNN stage which isolates features of the image. Based on the features selected, it is necessary to classify them as covid and non-covid cases. The severity of the patient will decide the line of treatment. SVM is a supervised classifier, hence it can be trained in less steps as compared to Random Forest or NN [16]. It can get maximum marginal hyperplane to classify the elements in the proper class. By iterating the hyperplane it finds the best distinctive classes.

12.4 RESULT AND ANALYSIS

As explained above in the proposed section, pre-processing of data and data augmentation are carried out on the COVID-19 X-ray database. It is used

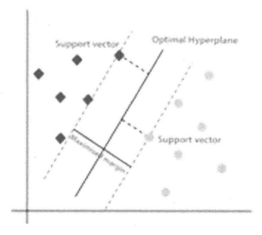

Figure 12.5 SVM classifier.

along with Python as the main programing language. The experiment is carried out on HP Elite book with a Core i5 processor.

12.5 CONCLUSION

The main aim of the research is to find a fast method to detect covid and non-covid cases. The conventional testing of RT-PCR and then X-ray is time-consuming and leads to delayed treatment for the severe patient. In our study, X-ray images itself work as input for the detection of COVID-19 patients and their severity. The images used for training of original dataset were less. So we have employed the image augmentation technique to increase the quality and size of the dataset. Such extended dataset help in training and increasing the performance of the model. CNN and SVM stages in cascade helped further to increase the speed and accuracy of the proposed model. The hybrid model takes advantage of deep learning plus the flexible hyperplane classification of the SVM. This hybrid approach is faster and more accurate than other computational methods. Our hybrid model approach will help practitioners to treat the patient early and that will lead to less rate of mortality in COVID-19. The fully integrated approach may help practitioners to treat patients as single-stage diagnostic at hospitals.

REFERENCES

[1] Garlapati, K., Kota, N., & Mondreti, Y. S. (2021). Detection of COVID-19 using X-ray image classification. In *Proceedings of the Fifth International Conference on Trends in Electronics and Informatics (ICOEI)*. USA.

[2] Hany, N., Atef, N., & Mostafa, N. (2021). Detection COVID-19 using machine learning from blood tests. In *International Mobile, Intelligent, and Ubiquitous Computing Conference (MIUCC)*. USA.

[3] Koushik, C., & Bhattacharjee, R. (2021). Symptoms based early clinical diagnosis of COVID-19 cases using hybrid and ensemble machine learning techniques. In *5th International Conference on Computer, Communication and Signal Processing (ICCCSP - 2021)*. USA.

[4] Oliveira, R. F. A. P., & Bastos-Filho, C. J. A. (6 June 2021). Machine learning applied in SARS-CoV-2 COVID 19 screening using clinical analysis parameters. *IEEE Latin America Transactions, 19*.

[5] Rahim, A., & Kurniawan, M. Machine learning based decision support system for determining the priority of covid-19 patients. In *2020 3rd International Conference on Information and Communications Technology (ICOIACT)*. USA.

[6] Miah, M. B., & Yousuf, M. (2015). Detection of lung cancer from CT image using image processing and neural network. *Journal of Big Data*. 10.1109/ICEEICT.2015.7307530.

[7] Reshi, A. A., Rustam, F., Mehmood, A., Alhossan, A., Alrabiah, Z., Ahmad, A., Alsuwailem, H., & Choi, G. S. (September 2021). An efficient CNN model for COVID-19 disease detection based on X-Ray image classification. *Mathematical Problems in Engineering*.

[8] Shorten, C., & Khoshgoftaar, T. M. (2019). A survey on image data augmentation for deep learning. *Journal of Big Data, 6,* 60.

[9] Dong, D., Tang, Z., Wang, S. et al. (2020). The role of imaging in the detection and management of COVID-19: a review. *IEEE Reviews in Biomedical Engineering, 14,* 16–19.

[10] Loddo, A., Pili, F., & Di Ruberto, C. (2021). Deep learning for COVID-19 diagnosis from CT images. *Applied Science, 23.*

[11] Roberts, M., Driggs, D., Thorpe, M., Gilbey, J., Yeung, M., Ursprung, S., Aviles-Rivero, A. I., Etmann, C., McCague, C., Beer, L., et al. (2021). Common pitfalls and recommendations for using machine learning to detect and prognosticate for COVID-19 using chest radiographs and CT scans. *Nature Machine Intelligence, 3,* 199–217.

[12] Canayaz, M. (2021). C+EffxNet: A novel hybrid approach for COVID-19 diagnosis on CT images based on CBAM and EfficientNet. *Journal of Chaos, Solitons & Fractals, 151,* 230–251.

[13] Yildirim, M., & Cinar, A. (2020). A deep learning based hybrid approach for COVID-19 disease detections. *Traitement du Signal, 37*(3), 461–468.

[14] Cohen, J. P. (2020). Github Covid19 X-ray dataset. https://github.com/ieee8023/covid-chestxray-dataset, 2020. Online.

[15] Isaac, A., Nehemiah, H. K., Isaac, A., & Kannan, A. (September 2020). Computer-aided diagnosis system for diagnosis of pulmonary emphysema using bio-inspired algorithms. *Computer in Biology and Medicine, 124,* 190–210.

[16] Oulefki, A., Agaian, S., Trongtirakul, T., & Laouar, A. K. (June 2021). Automatic COVID-19 lung infected region segmentation and measurement using CT-scans images. *Pattern Recognition, 114,* 1120–1135.

Chapter 13

Deep Learning Algorithms for IoT-Based Hybrid Sensor Using Big Data Analytics

R. Ganesh Babu, S. Markkandan, A. Sugunapriya, and
P. Karthika

CONTENTS

13.1 INTRODUCTION

Assembling plays an important role in monetary advancement and is still viewed as critical to financial development in the age of globalization [1]. It has a positive impact on the growth of both created and created nations. The development of advances is used by the assembly business to improve the financial intensity of the individual producers and the mechanical sustainability of the whole segment. Data and communication invention (ICT) collection in assembly enables development from tradition to cutting-edge ways of manufacturing [2]. Checking systems as a major aspect of the ICT application has a major impact on integrating process control and managers.

DOI: 10.1201/9781003355960-13

Continuous advances in data innovation empower the mix of various control applications into one complex framework for the entire inventory network.

Quality expectations and better dynamic support to directors with the expanding number of accessible IoT detecting gadgets, information produce since the assembly business is used to developed [3]. This kind of information is classified as "big data." Huge information review has prompted critical improvements in assembly operations, such as decreasing the use of vitality, improving creation bookings and arranging coordination, moderating social hazards, and encouraging better dynamics [4].

The past investigations showed enormous interest in incorporating huge advances in information, reducing the preparation occasion for home robotization frameworks, as long as successful and effective responses to the preparation of IoT-produced information for brilliant urban communities, and dealing with IoT-shrewd environmental information on an ongoing basis in Figure 13.1.

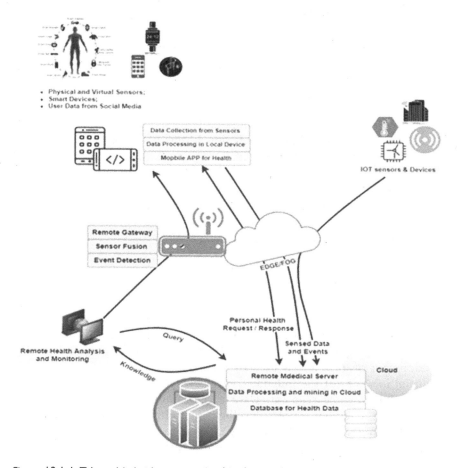

Figure 13.1 IoT-based hybrid sensor using big data analytics.

The consequences of the above-mentioned examinations have demonstrated critical IoT-based sensor focal points, large-scale information innovation, and AI models in improving executive dynamics [5]. In any case, no examination of the reconciliation of IoT-based sensors, large-scale information innovation, and AI models into a total observation framework for car manufacturing explicitly. In the proposed continuous testing system that uses IoT-based sensors, huge preparation of knowledge and a model of cross-breed forecasting for the car business [6]. The projected IoT-based sensor gathers temperature, wetness, meter, and gyrator information from the phase of the mechanical production system, while the enormous information planning the stage manages and stores other items sensor in sequence [7].

Finally, the proposed half-breed forecast model, consisting of DBSCAN-based exception discovery and Random Forest order, is used separately to evacuate anomaly information and identify deficiencies during assembly.

13.2 IOT SENSOR OBSERVE SYSTEM

Late innovations of IoT, sensors, huge information, and AI used for observation can take on significant jobs in anticipating illness, improving creation, reducing expenditure, providing an early warning framework, and encouraging better dynamics for executives. A few reviews were conducted on IoT-based observation frameworks and notable favourable circumstances were indicated. The proposed IoT-based system would check fundamental human signs. A contextual analysis was conducted regarding observation of the pulses of footballers during a football coordinate. The proposed framework had the option of screening and anticipating the most noticeably terrible circumstances of the player's imperative signs (i.e. unexpected demise), but also potential wounds [8].

The proposed model is an observational framework for IoT-dependent rural fields. The created framework was used to dynamically screen the dampness of citrus soil and treatment supplements and the water system [9]. The results of the contextual investigation indicated that the framework proposed helped the ranchers to make better choices. The proposed framework was compelling to observe the earth and issue a notice acknowledging basic opportunities.

Finally, projected continuous observation is to monitor the well-being of construction destinations depending on data display and a remote sensor system. Remote sensor hubs are collected and shipped to a remote server. Once an irregular circumstance has been identified, a notice/caution was issued by the proposed framework in Figure 13.2. Contextual survey results indicated that the projected framework enhanced the safety of the building administration progressively with better dynamics [10].

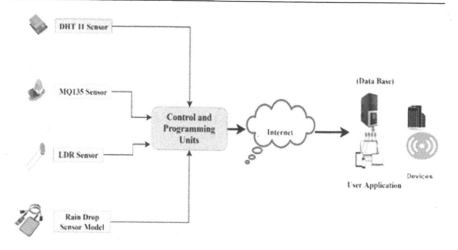

Figure 13.2 IoT-based sensor monitoring system.

In a recreation situation, the proposed framework was implemented. The results indicated that the use of a few IoT-based sensors within a brilliant structure would achieve a superior test system. The proposed framework is needed to improve vitality productivity in the same way as encouraging green shrewd structures [11]. The proposed IoT-based sensor for inspecting radon gas levels within a structure. The proposed framework could tell / caution customers when it comes to preventing risky circumstances for a given degree of radon gas. The proposed framework had the option of screening the radon gas level, triggering the modified activities, and telling customers once a certain degree of radon gas had arrived. The proposed framework for secluded indoor air quality observation collects a few sorts of sensor information, such as CO_2, CO, SO_2, NO_2, O_3, Cl_2, temperature, and wetness. The sensor information was processed by using a solitary board PC (Raspberry Pi) as a door [12].

The proposed natural observation framework is based on the ease of IoT sensors for preventing blunders in added substance producing during the structure stage [13]. The sensors were used to assemble the information regarding temperature and stickiness. The examination found that information on natural conditions could help prevent blunders in additional substance manufacturing during the structural phase. It utilized IoT sensors to gather information for the mine lifting gear issue conclusion. The investigation revealed that IoT sensors will help give complete information about the conclusion just as they boost the results of the decision. The system proposed using IoT and AI to anticipate the nature of an item and to advance control of activity. Metal giving used a role as a valid case execution of the projected framework.

The IoT-based quantity of sensors and associated segments is fundamentally expanding. IoT's appropriation in assembling enables the progress from customary digitized production to the present day [14].

13.3 BIG DATA DISPENSATION

With the growing number of IoTs and the detection of gadgets, information generated from the assembly of frameworks is used to develop exponentially, supplying alleged "enormous information." In terms of information age speed, the third V is speed and the last V is actuality in terms of the unwavering quality of the information. The information produced during assembly expands with different sorts and arrangements day by day, henceforth the preparation and capacity of this information become a difficult issue that should be targeted. In the assembly business, there are a few uses of huge investigation of information. In a concentrated assembly industry, it projected a major information structure to decrease the use of vitality and discharge. The proposed framework consists of two segments, which secure information to gather vitality information and examine information to investigate vitality utilization. The results showed that, given the real use of the case, the proposed framework was equipped to kill 4% of the vitality utilization and 5% of the vitality costs. It projected a major information framework for the disclosure of mining information by the RFID-enabled creation information co-ordinations.

A trial is used to make obvious the potential of the projected framework and the outcome indicated to the information gathered from extensive information is used to create reservations and arrange coordination. It considered the use of massive information research to moderate social hazards in the network of inventories. A contextual investigation was used to broaden the use of massive review of information in the production network. The results of the review revealed that huge information research can help the board anticipate different social issues and alleviate social dangers. They proposed a major information system to assemble forms of complex occasions for dynamic detection and preparation. A connection model was created and XML-based assembly forms put together to successfully process large information on complex occasions. The *a priori* visit thing mining calculation was used to find a continuous example of the intricate occasions [15]. With execution in a neighbourhood bean stew sauce-producing organization, the attainability and appropriateness of the proposed framework were affirmed. The proposed model is based on giving the board dynamic direction to earth.

The proposed structure provides continuous information for the IoT occasions. The proposed structure used Apache Kafka as a framework for the message line and was able to process the information on continuous IoT occasions. It proposed a system which would rely on huge innovations in

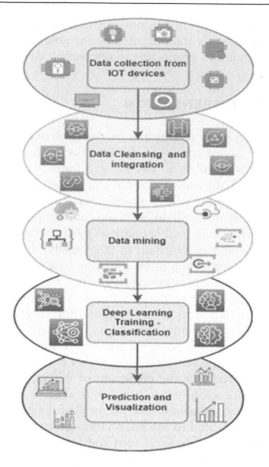

Figure 13.3 Big data processing for deep learning.

information and AI expectations for online Figure 13.3 wind turbine problems. Apache Kafka was continually used to deal with the approaching information and to throw the information to a spilling framework for additional investigation. The projected framework is used to screen wind turbine position and is needed to assist lower activity and boarding costs. It proposed a structure that would take care of colossal measures to approach unstructured vehicle (CV) information. The proposed structure utilized Apache Kafka as a distributor of messages. Exploratory results showed that the proposed framework is sufficiently effective to address tremendous measures to address CV information and achieve the estimated differentiation suggested by negligible dormancy. In the assembly sector, the engineering proposal for the machine signs for an ingestion framework is dependent on Apache Kafka. The projected system collects computer logs of a large number of processing devices, manages them in an information line and passes them on to external systems for further investigation.

Finally, the board framework in an assembly shop floor relies on huge information innovations for machine-generated information. The proposed framework uses Apache Kafka as a communication to the Apache Storm as a framework for continuous preparation. The proposed framework must be implemented to reduce foundation and organization costs.

Apache Storm is an equivalent framework that continues to circulate for preparing information on high-speed streams. It is tolerant and adaptable to the problem, with assured handling of the information. Apache Storm had been used in past tests to continuously prepare data. This proposed a stream-based system that would give continuous data advantages over open travel. The proposed system utilized Apache Storm as a motor for handling continuous circulation. Results indicated that the structure proposed was fit to take care of lots of constant information with lower inactivity [16]. The expanding measure of sensor information produced by IoT has prompted increased interest in sensor-accommodating stages in the storage of information. Due to their evolving adaptability, versatility and accessibility NoSQL databases have become well known over the last few years. All things considered, the term "NoSQL" alludes to information storage phases that do not follow a severe information reproduction for public databases. The database archive that offers adaptable patterns of information, superior quality, versatility and accessibility. A previous report looked at the MongoDB and Oracle exhibition with tests added, updated, and erased. In all tests, MongoDB outlines the prophet. Similarly, MongoDB has been shown to be powerful in putting away inventory network information, geographic data frameworks, and assembling. MongoDB stores sensor information created by IoT to observe a transitory inventory network of nutrients. Similarly, MongoDB has been shown to be powerful in putting away inventory network information, geographic data frameworks, and assembling. They managed a near-report among six well-known databases to look after an assortment of geospatial information. The results illustrate that MongoDB was satisfactory in conditions of equality of inquiry and utilization of assets for taking care of spatiotemporal information; it projected MongoSOS, a sensor perception administration reliant on MongoDB.

The proposed framework was equipped to take care of the peruse and compose access to the route and locate information while the display was enhanced by approximately 2% and the normal model. It proposed a major information handling framework dependent on Apache Spark and MongoDB to recognize the gainful zones from a lot of information about taxi trips. The test results indicated that the proposed framework was adaptable and sufficiently competent to prepare gainful territory questions from tremendous measures of huge information on taxi trips. It proposed a NoSQLMongoDB-dependent, adaptable information diagram for the virtualization of assembly machines. The proposed framework was assessed against a few articulations of the investigation.

13.4 AI METHODS IN MANUFACTURING

In the information age, the assembly business is experiencing an expansion, e.g., creation line sensor information, ecological information, etc. New innovation advances, for instance, AI offers incredible potential to break down information archives and can then provide dynamic management assistance or be used to improve the implementation of the framework [17]. AI procedures are used to identify some examples or regularities and have been effectively executed in different territories, such as location of deficiencies, quality expectations, deformity order and visual examination. A few investigations used AI and proved to be critical in assembly operations. To identify oddities, it employed seven distinctive AI-based techniques for injured wafers. The models were prepared to use Fault Detection and Classification (FDC) information to identify defective. The test results illustrate a high probability of distinguishing flawed wafers from AI-based models. The evaluation study conducted four AI calculations to anticipate the character of the metal castings item. The result showed that all four AI calculations can be used to provide a viable way of anticipating the nature of the item. It used help vector calculator to predict welding in a powerful layer of plates. The results indicated that the projected quality forecasting reproduction could be utilized for the ongoing framework of observations. To limit the erroneous admonition in distinguishing the nature of the assembled item, a wise framework was created. Real use in assembling organizations can be adequate within the proposed framework. It is used to limit the erroneous arrangements and to improve the value estimate display. Finally, two AI calculations were also used to mechanize the way in which the nature of machine parts was investigated. The quality of three kinds of machine parts was estimated.

Short identification and finding is a significant issue in process building and is used in a procedure to recognize anomalous occasions. Early recognition of procedural shortcomings can help to avoid profitability misfortunes. In Figure 13.4, AI calculations showed enormous adequacy in identifying process blames when assembling. Arbitrary Forest is an outfit forecast strategy which sums up the after-effects of the individual trees of choice. In general, Random Forest works to create subsets of information preparation by using the sacking strategy. A calculation of the tree of choice is used for each preparation dataset. Finally, the final outcome of the forecast is chosen depending on the dominant part vote over all the trees in the timberland. As of late used Random Forest to recognize the rotor bar's disappointment. They played out the investigation of the presentation between the Random Forest and various models. The test results illustrate that Random Forest beat different representations and has a precision of approximately 98.8%. Just as the framework for preventive support in industrial facilities, the proposed model can be used for continuous deficiency observation framework.

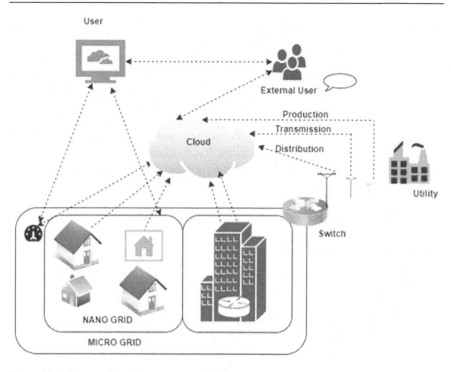

Figure 13.4 AI manufacturing connected devices.

The results were contrasted and those were obtained from a current method of man-made brainpower, the neural system. The results showed that execution of Random Forest would be advised and higher accuracy than calculation of the neural system. The consequences of this investigation are normal for the discovery and analysis of bearing problems. Lastly, proposed weakness finding in prod gears depending on hereditary calculation and Random Forest. The projected framework sections to be specific inherited calculation for quality determination and grouping of Random Forest. The proposed framework has been tried on genuine signs of vibration and Random Forest would be wise to implement for lack of finding.

The calculation finally recognizes focuses that have no place with any bunch, that are treated as exceptions. DBSCAN was executed in various territories and, by recognizing genuine exceptions, indicated critical precision. It suggested an exceptional location strategy including a delicate sensor that demonstrates time schedule. They used DBSCAN for anomaly recognition and exhibited great execution in the proposed exception discovery strategy. It proposed the DBSCAN-dependent exception recognition for sensor information in remote sensor systems. The proposed model effectively isolates exceptions from ordinary information on sensors. The proposed model demonstrated noteworthy precision in identifying exceptions in the light of analyzes with engineered datasets, among an accuracy rate of 99.5%. In addition, a few reviews indicated enormous results for anomaly location based on DBSCAN with respect to improving the accuracy of the arrangement [18]. To propose a cross-breed expectation model consisting of an exceptional location based on DBSCAN to evacuate the anomaly information, and Random Forest to determine whether the assembly procedure works typically or anomalously. The crossover expectation model is combined with an ongoing enormous knowledge planning system, enabling the handling of sensor information from the IoT-based sensor gadget and continuous fault prediction.

13.5 METHODOLOGY

13.5.1 System Design

The proposed constant observing framework was created here to assist chiefs with bettering screen the mechanical production system process in a car fabricating process just as gives early notice when a deficiency is distinguished [19]. The proposed framework utilizes IoT-based sensors, handling huge information and a model of crossover forecast. The half-and-half forecast model comprises of a bunched external location framework. As can be seen in Figure 13.5, the sequential construction system connects sensors to the work area of a workplace. The sensor information produced by IoT is transmitted remotely to a cloud server on which the large information handling framework is introduced. The framework permits the framework to rapidly process a lot of sensor information before it's put away in the MongoDB database. To sift through the exceptions from the sensor information, a grouping-based anomaly recognition strategy is utilized. Also, an information examination machine-based learning grouping model is utilized to anticipate blames in current sensor information during the mechanical production system process. At long last, the total history of the sensor information, for example, temperature, stickiness, accelerometer, and gyrator information is introduced to the administrator continuously by means of an online observing framework notwithstanding the after-effects of the flaw forecast.

Data mix and sharing accept an essential activity in making unavoidable circumstances subject to IoT data. This activity is progressively fundamental for time-delicate IoT applications where a promising blend of data is relied upon to bring all bits of data together for examination and subsequently give trustworthy and careful essential bits of information. They presented a review paper in which data blend frameworks for IoT conditions are investigated trailed by a couple of possibilities and troubles [20].

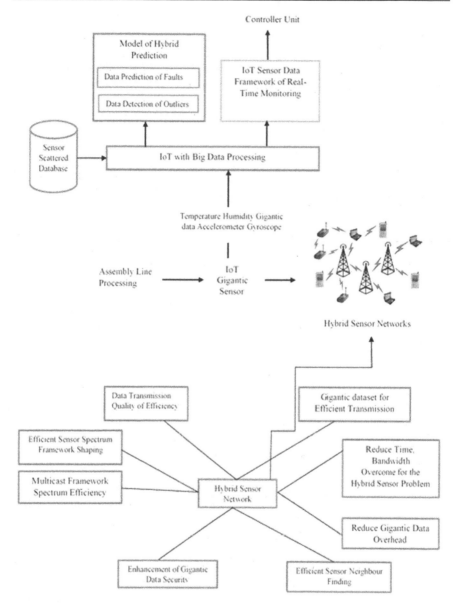

Figure 13.5 Hybrid sensor architecture.

13.5.2 IoT Sensor Data

Many research attempts proposed spilling data examination that can be generally sent on unrivalled figuring systems or cloud stages. The spilling data assessment on such structures relies upon data parallelism and continuous dealing with. By data parallelism, a gigantic dataset is distributed to a couple of more diminutive datasets, on which equivalent examinations are played out at the same time. Slow planning insinuates carrying a little gathering of data to be dealt with quickly in a pipeline of estimation assignments. Regardless of the way that these frameworks decrease time dormancy to reestablish a response from the spilling data explanatory structure, they are not the best response for time-stringent IoT applications. By bringing spilling data assessment closer to the wellspring of data (i.e. IoT devices or edge contraptions), the pre-requisite for data parallelism and consistently taking care of is less sensible as the size of the data in the source grants it to be arranged rapidly. Regardless, expediting snappy examination IoT devices introduces its own troubles, for instance, imperative of figuring, accumulating, and power resources at the wellspring of data.

13.5.3 IoT with Big data

IoT is prominent to be one of the noteworthy wellsprings of gigantic data, as it relies upon partnering a massive number of savvy devices to the Internet to report their routinely got status of their environment. Seeing and expelling critical models from huge rough data is the middle utility of immense data assessment as it realizes increasingly raised degrees of bits of information for essential administration and example desire. Right now, these bits of information and data from the enormous data are of silly hugeness to various associations, since it engages them to build advantages. In human sciences, Hilbert ponders the impact of enormous data assessment to that of the development of the telescope and amplifying focal point for space science and science, independently. A couple of works have depicted the general features of gigantic data from different viewpoints with respect to volume, speed, and arrangement.

13.5.4 Big Data with Hybrid Sensor Network

Standard data use bound together database structure in which gigantic and complex issues are enlightened by a single PC system. United building is costly and inadequate to process tremendous proportion of data. Gigantic data relies upon the scattered database design where a colossal square of data is fathomed by apportioning it into a couple of more diminutive sizes. By then, the response for an issue is prepared by a couple of unmistakable PCs present in a given PC mastermind. The PCs confer to each other in order to find the response for an issue. The coursed database gives better figuring, lower cost and besides improve the introduction when appeared differently in relation to the united database system. This is in light of the fact that the focused plan relies upon concentrated PCs which are not as money related as microchips in passed-on database systems. In a similar manner, the appropriated database has progressively computational power when appeared differently in relation to the consolidated database system which is used to manage traditional data.

The implanting plan is appropriate for an enormous storehouse of sensor information which requires quick peruse and compose yield. In this manner utilized an implanting plan-based information vault for sensors in our investigation. The sensor archive comprises of IoT gadget ID, documentation time, handled time with sensor information, also expectation outcome as can be seen. The sensor information, for example, information on temperature, stickiness, spinner and accelerometer are consolidated as a sub-report.

13.6 IMPLEMENTATION

The checking framework was applied in this investigation to screen the sequential construction system process for entryway trim creation at a car fabricating plant in Figure 13.6. The created IoT-based sensor comprises of a solitary principle board, Raspberry Pi and Sense-HAT as an extra sensor board. Raspberry Pi is a little single-board gadget with measurements of 85.60 m, 53.98 mm, and 17 mm, weighing only 45 g, and is cheap at around $25–35 USD. It has USB, LAN, HDMI, sound, and video associations for different information and yield activities. Likewise, broadly

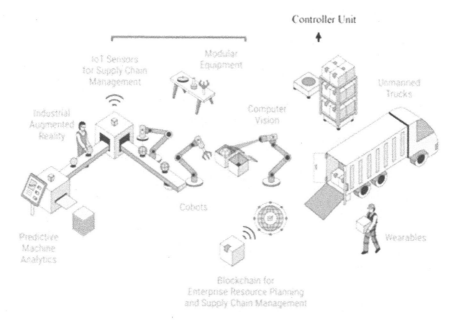

Figure 13.6 Real-case implementation within an assembly line of the proposed IoT-based sensor.

useful info yield (GPIO) connectors permit extra gadgets to be associated with the principle board, or extra sheets, for example, sensors. The Sense-HAT board is an extra sensor board which quantifies the temperature, mugginess, and accelerometer and spinner information and is intended for the Raspberry Pi as an official extra board. The Sense-HAT board is attached by means of GPIO 40 pins to a Raspberry Pi. The IoT-based sensor gadget forms amassed and executed in genuine cases.

13.6.1 Half-and-Half Prediction Model for Fault Detection

Half-and-half expectation model is being used right now to determine whether the procedure works normally or anomalously. Figure 8.4 shows the way in which to classify specific or unusual occasions during the assembly process. The crossover expectation model uses a DBSCAN-dependent anomaly position to identify and expel sensor information anomalies to ordinary and unusual occasions. Finally, the presentation is evaluated by contrasting the half-and-half predicted model with other order models.

13.7 RESULTS AND DISCUSSIONS

The representation of information was generated by using JavaScript structure as an observing system to gradually view sensor information. The administrator should screen the status of the sequential construction system process as soon as the strange opportunity (deficiency) is gradually differentiated through the proposed framework. As an Internet browser on a PC can be used to handle the ongoing observing process. The projected system incorporates the sensor details, such as knowledge about temperature, stickiness, and accelerometer with gyrator gradually. For each log, the gadget ID (IoT-based of sensor gadget), and reported time were composed and displayed. What's more, the expectation model for the mixture was used to predict the deficiency and bring the outcome into the ongoing testing process. The proposed framework was updated and tested in one of the manufacturing of automobiles. In the assembly sequential construction system, four IoT-based sensor gadgets were introduced and each 5 s transmitted the sensor information to the remote server. Around 19 million records have been collected during this testing period. Our proposed constant checking framework of three sections: the IoT-based sensor, the large stage of information handling, and the model of half-and-half expectations.

13.7.1 IoT-based Sensor Execution

The IoT-based system of the sensor gadget and a customer software to collect and transmit sensor information to a cloud server. The breakdown of the IoT-based execution of sensors under different conditions is critical. Measurements of execution, for example, organize deferral and CPU, and use of memory right now. This proposed to organize delay as a measure to evaluate the execution of the sensor gadget while using CPU as a measure to evaluate the capacities of IoT gadgets in different situations. In our exam organize impediment was characterized as the normal time source sending the sensor information. The second execution metric was the customer program's normal CPU and memory usage under various situations.

13.7.2 Big Data Processing Performances

Dissecting the exhibition of enormous handling of information under different conditions is important. Measurements of execution, such as frame dormancy, throughput, and simultaneousness, have been used right now. Framework inertness is characterized in our investigation as it time required through the proposed framework to deal with, process, as well as a store of sensor information in a database. The final metric is simultaneity that is clear the number of customers having entered the system simultaneously. The experiments were performed with various server numbers, and the reaction time was obtained for investigation.

Figure 13.7a demonstrates that, since the assessment of sensor information shipped off the cloud worker has improved, the reaction time has moreover extended. The amount of customers in a similar manner affected the reaction time as the proposed system required a more prominent ability to measure and store sensor information sent by an undeniably obvious number of customers continually. While more workers are consolidated, manhandling the adaptability backing will help achieve lower reaction time when diverged from a particular worker as showed up in Figure 13.7b. Figures 13.7c and 13.7d show the contraption throughput with different number of customers. Improving worker numbers could acquire better execution. Figures 13.7e and 13.7f mull over latency and information base of MongoDB and CouchDB contraption size.

13.7.3 Hybrid Fault Prediction Model

During the dataset age, the enormous data dealing with structure get the sensor data from the IoT-based sensor unit and stores the data. The IoT-based sensor gathers data from various kinds of activity, including normal and odd chances. By then, ace customers name the dataset relying upon the status of the strategy during when the sensor data is given to be assembled. Next, the dataset is analyzed utilizing the anticipated blend model to foresee the status of issue. The discoveries of the plan affiliation are fused in a couple of ordinary characterization models for a couple of course of action models. For model, with perceiving and envisioning abnormal events, Naïve Bayes (NB), Logistic Regression (LR), Multilayer Perceptron (MLP), and

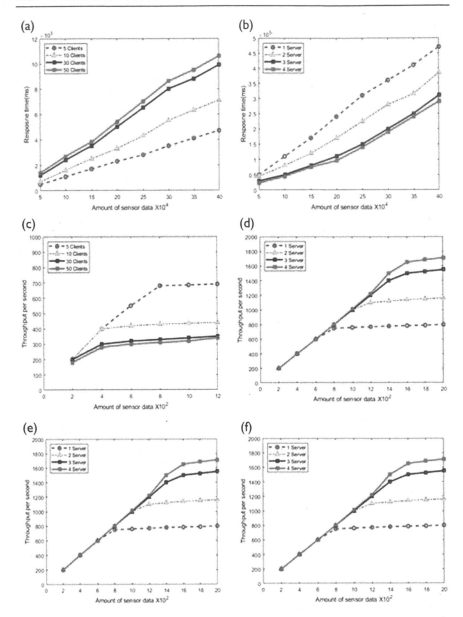

Figure 13.7 Assessment of latency performance with different client/server numbers (a) response time of client numbers and (b) response time of server numbers; (c) throughput with different client numbers; (d) throughput with different server numbers; (e) response time between MongoDB and CouchDB databases; and (f) database size between MongoDB and CouchDB.

Table 13.1 Execution correlation of several classification models for fault prediction

Model	Precision (%)	Recall (%)	Accuracy (%)
Naive Bayes (NB)	94.1	93.6	93.567
Logistics regression (LR)	98	98	97.967
Multilayer perception (MLP)	96.9	96.9	96.879
Random Forest (RF)	98.6	98.6	98.789
DBSCAN+NB	96.8	96.7	96.76
DBSCAN+LR	98.7	98.6	98.67
DBSCAN+MLP	98.9	98.9	98.81
Hybrid prediction model (DBSCAN+RF)	100	100	100

Random Forest (RF) appeared differently in relation to the cross-variety desire model. The model proposed accomplished the most noteworthy precision (100%) comparative with other structure models. The proposed model was done in Apache Storm where Kafka's floods of data can be orchestrated and foreseen in a specific and industrious way. The deferred outcomes of usage where Apache Storm continues anticipating whether the strategy works regularly or phenomenally given the IoT-based sensor subtleties (e.g. temperature, sogginess, accelerometer, and whirligig). The delayed consequences of Table 13.1 the review are needed to help overseers in hindering unexpected misfortunes achieved by lacks close to the start and to help components during the party method.

13.8 CONCLUSION

At the present time, a progressing testing framework has been built up that utilizes IoT-based sensors, getting ready a lot of data, and a model of cream desires. The proposed model is used to help bosses in checking the status of the mechanical creation structure measure and to see charges continually, while unforeseen episodes achieved by needs can be forested accordingly. Through this assessment, we indicated that it is productive to encourage IoT-based sensors with a critical information dealing with framework for reliably preparing and investigating a huge load of sensor information. Apache Kafka, Apache Storm, and NoSQLMongoDB made the tremendous information managing framework right now. The exploratory outcomes showed that the framework is adaptable and can deal with a lot of steady sensor information more competently than standard models. In addition, the IoT-based sensor show was isolated with various assessments, for instance, system deferral, CPU and memory utilization. For each test condition, the IoT-based sensor gave a productive strategy as it effectively gathered and sent the information inside an adequate time with low computational cost. Shortcoming affirmation is a basic issue in the accumulating strategy as it can see whether the system is working ordinarily or unusually. The proposed half-and-half craving model including peculiarity region subject to DBSCAN and demand from Random Forest. DBSCAN was used to withdraw uncommon cases from standard sensor information, while Random Forest was used to envision sensor information insufficiencies. The results indicated that the proposed cross-variety estimate model is incredible in standing out high exactness from various models attempted. The discoveries of the investigation are utilized to help chiefs and lift elements during get-together, maintaining a strategic distance from unintended disasters brought about by shortcomings. Well-being is a significant issue when acquiring, leading, and partner further IoT devices. From that point, the strength of IoT gadgets and stages ought to be considered in a future report. What's more, a variety of odd conditions ought to be distinguished and gathered during the get-together procedure with the goal that the proposed cross-variety desire model can be utilized in the near future to profit by a complex dataset.

REFERENCES

[1] Rashid, M., Singh, H., Goyal, V., Parah, S. A., & Wani, A. R. (2021). Big data based hybrid machine learning model for improving performance of medical Internet of Things data in healthcare systems. In *Healthcare Paradigms in the Internet of Things Ecosystem* (pp. 47–62). Academic Press.

[2] Lin, S. Y., Du, Y., Ko, P. C., Wu, T. J., Ho, P. T., & Sivakumar, V. (2020). Fog computing based hybrid deep learning framework in effective inspection system for smart manufacturing. *Computer Communications, 160*, 636–642.

[3] Sampathkumar, A., Maheswar, R., Harshavardhanan, P., Murugan, S., Jayarajan, P., & Sivasankaran, V. (2020, July). Majority voting based hybrid ensemble classification approach for predicting parking availability in smart city based on IoT. In *2020 11th International Conference on Computing, Communication and Networking Technologies (ICCCNT)* (pp. 1–8). IEEE.

[4] Doulamis, N. (2018). Adaptable deep learning structures for object labeling/tracking under dynamic visual environments. *Multimedia Tools and Applications, 77*, 9651–9689.

[5] Li, W., Chai, Y., Khan, F., Jan, S. R. U., Verma, S., Menon, V. G., & Li, X. (2021). A comprehensive survey on machine learning-based big data analytics for IoT-enabled smart healthcare system. *Mobile Networks and Applications,. 26*(1), 234–252.

[6] Amanullah, M. A., Habeeb, R. A. A., Nasaruddin, F. H., Gani, A., Ahmed, E., Nainar, A. S. M., Akim, N. M., & Imran, M. (2020). Deep learning and big data technologies for IoT security. *Computer Communications, 151*, 495–517.

[7] Yan, H., Wan, J., Zhang, C., Tang, S., Hua, Q., & Wang, Z. (2018). Industrial big data analytics for prediction of remaining useful life based on deep learning. *IEEE Access, 6*, 17190–17197.

[8] Shakarami, A., Shahidinejad, A., & Ghobaei-Arani, M. (2021). An autonomous computation offloading strategy in mobile edge computing: A deep learning-based hybrid approach. *Journal of Network and Computer Applications, 178*, 102974.

[9] Yousefi, S., Derakhshan, F., & Karimipour, H. (2020). Applications of big data analytics and machine learning in the internet of things. In *Handbook of Big Data Privacy* (pp. 77–108). Springer, Cham.

[10] Ghorpade, S. N., Zennaro, M., & Chaudhari, B. S. (2021). IoT-based hybrid optimized fuzzy threshold ELM model for localization of elderly persons. *Expert Systems with Applications, 184*, 115500.

[11] Machorro-Cano, I., Alor-Hernández, G., Paredes-Valverde, M. A., Rodríguez-Mazahua, L., Sánchez-Cervantes, J. L., & Olmedo-Aguirre, J. O. (2020). HEMS-IoT: a big data and machine learning-based smart home system for energy saving. *Energies, 13*(5), 1097.

[12] Khan, P. W., Byun, Y. C., & Park, N. (2020). IoT-blockchain enabled optimized provenance system for food industry 4.0 using advanced deep learning. *Sensors, 20*(10), 2990.

[13] Ganesh Babu, R., & Amudha, V. (2018). A survey on artificial intelligence techniques in cognitive radio networks. In *Emerging technologies in Data Mining and Information Security. Advances Intelligent Systems and Computing, 755*, 99–110.

[14] Ganesh Babu, R., Karthika, P., & AravindaRajan, V. (2020). Secure IoT systems using Raspberry Pi machine learning artificial intelligence. In *ICCNCT 2019: Second Internationational Conference on Computer Networks and Communication Technologies, Lecture Notes on Data Engineering and Communication Technologies, 44*, 797–805.

[15] Ganesh Babu, R., Karthika, P., & Elangovan, K. (2019). Performance analysis for image security using SVM and ANN classification techniques. *Proceedings of Third IEEE International Conference on Electronics, Communications and Aerospace Technology (ICECA)*, Vol. 12, 460–465.

[16] Mittal, S., & Sangwan, O. P. (2019). Big data analytics using machine learning techniques. *Proceedings of 9th IEEE International Conference on Cloud Computing, Data Science and Engineering (CONFLUENCE)*, Vol. 112, 203–207.

[17] Nedumaran, A., Ganesh Babu, R., MeseleKass, M., & Karthika, P. (2019). Machine level classification using support vector machine. *AIP Conference Proceedings of ICSMMT*, Coimbatore, India, *2207*, 020013-1-020013-10.

[18] Pouyanfar, S., Sadiq, S., Iyengars, S. S., & Yan, Y. (2018). A survey on deep learning: algorithms, techniques, and Applications. *ACM Computing Surveys, 51*, 1–36.

[19] Tsai, C., Lai, C., Chao, H., & Vasilakos, A. V. (2015). Big data analytics: a survey. *Journal of Big Data, 2*, 1–32.

[20] Youssra, R., & Sara, R. (2018). Big data and big data analytics: Concepts, types and technologies. *International Journal of Research and Engineering, 5*, 524–528.

Chapter 14

Delineation of Landslide Hazard Zones of a Part of Sutlej Basin in Himachal Pradesh Through Frequency Ratio Model

Asutosh Goswami, Suhel Sen, and Rupa Sanyal

CONTENTS

14.1 INTRODUCTION

The phenomenon of sudden downward movement of rock debris along the slope is known as landslide [1]. Landslides are the manifestations of not only geological, hydrological, and geomorphological conditions but also some other factors, namely, rainfall, earthquake, land use changes, etc., are responsible for creating high-intensity landslides [2]. The terrain instability along with slope failure trigger mass wasting of various intensities, which causes not only the infrastructural but also the environmental damage [2]. Landslides are basically natural phenomena, but it turns into a devastating hazard and causes loss of life and property. According to De [3], human activities such as uncontrolled deforestation, unscientific land use policy, development of towns and tourism industry at the cost of mountain biodiversity, etc., are responsible for the intensification of landslide vulnerability of an area. Rainfall is also considered as a triggering factor for high-intensity landslides [4]. Biswas and Paul [4] also identified some of the triggering factors for the generation of landslides such as earthquakes, neotectonic activity, etc. Most of the hilly tracts of India is marked by the occurrence of the landslide of various intensities as the steep slopes are more prone to rock fall and debris fall in comparison to its counter gentle slope. The change of slope from a steady to an unsteady state is also found to be a significant cause of landslides [5]. It is also considered as a potential natural disaster, which causes the destruction of life, property and other material resources more particularly in the mountainous areas [6]. The devastating impacts of landslide events have well been established for the loss of human lives to the loss of valuable resources [7]. The landslide phenomena cannot be avoided but the identification of potential areas of landslide can reduce the vulnerability of landslide to some extent. Wireless sensor networks have vast applications in the monitoring of natural hazards or disasters such as landslide [7].

The softer rock groups have a higher potentiality to face landslide more because of their lesser shear strength [8]. Pardeshi et al. [9] identified landslide as a significant geological hazard. Landslide is also considered to be the most frequent natural hazard that affects the livelihood of people worldwide [10].

The intensity and frequency of landslides are expected to increase in near future due to uncontrolled deforestation and haphazard town planning along with the climate changes in the landslide-prone areas [11]. The principal driving force for the landslide is the force of gravity. According to Chae et al. [12], landslide monitoring is an inevitable part of landslide assessment. In recent times, the landslide monitoring especially the early warning system has gained popularity. In reality, different techniques and sensors are involved in the process of monitoring landslide events [13]. With the invention of GIS and remote sensing technology, the preparation of landslide susceptibility zone (LSZ) has become easier [14]. Satellite remote sensing (RS) technology is an important step in this regard, as it can monitor the hazards and disasters for a wider area [15].

Diverse RS data such as space-borne synthetic aperture radar (SAR) and optical remote sensing, airborne light detection and ranging (LiDAR), etc., are in use for the continuous monitoring of landslides [16]. According to

Brabb [17], 90% of the total landslide hazards around the world can be minimized through an efficient forecasting system. The landslide susceptibility mapping is very helpful for the detection of potential landslide vulnerability zones thus reducing the associated disaster risk and enhancing the mitigation plans [14]. Natural hazards such as earthquakes, landslides, etc., can cause enormous loss of capital and life in the mountain areas more specifically [15].

A number of studies have been made by researchers for the identification of LSZ. Fayez et al. [18] applied a statistical technique known as FRM for the delineation of LSZ in a part of the state of Uttarakhand in India.

Bin et al. [19] in their work made an attempt in preparation of landslide susceptibility map of Qinghai province of north-western China. Elmoulat and Brahim [20] made an attempt to prepare an LSZ using GIS and weights of evidence model in Tetouan-Ras-Mazari area located in northern Morocco. The study area has covered of about 590 km^2 of the western riff region and the said study has taken into consideration of 126 landslide events and 6 conditioning factors. However, for modelling purposes, only 50% of the samples have been utilized. The landslide susceptibility map (LSM) has been classified into four classes namely low, moderate, high and very high and is found to be 95.3% accurate. The susceptibility map reveals that highly susceptible areas of landslides are having outcrops of mudstones, marlstones, and conglomerates with higher elevation and slope of more than 57°.

Vojtekova and Vojtek [21] aimed for the preparation of LSM at a local spatial scale around the city of Handlova in Slovakia by employing analytical hierarchy process (AHP). Berhane et al. [22] aimed for the preparation of landslide susceptibility map of Adwa-Adigrat mountain chains of Northern Ethiopia using frequency ratio model (FRM) through the application of geospatial technique. The study revealed that out of 546 landslides events, about 73.8% of the landslides fall within the high and very high LSZs. Nohani et al. [23] in their work made an attempt to prepare the landslide susceptibility map of Klijanrestagh Basin of northern Iran using GIS-based bivariate models. However, all other three models were also considered to be reasonable and accurate. OziokoandIgwe [24] made an attempt to prepare a GIS-based landslide susceptibility map for Udi-Iva valley of southeast Nigeria by applying heuristic AHP and frequency ratio (FR). The landslide map prepared was divided into five zones namely VL, L, M, H, and VH susceptibility zones. Anis et al. [25] prepared an LSM of north-western Tunisia by using bivariate statistical technique. Ahmad et al. [26] in their work aimed for the preparation of landslide susceptibility map of central Zab basin in the western Azerbaijan province of Iran. The present study has the aim to delineate the LSZs of a part of the Sutlej basin of Himachal Pradesh

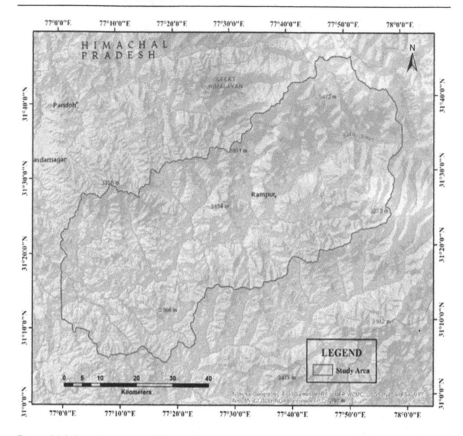

Figure 14.1 Location map of the study area.

state in India (Figure 14.1) by using geospatial techniques through the application of FRM which will help in undertaking proper planning strategies for the development of the people residing in the study area. The present study is also undertaken for predicting the occurrences of landslide events in the study area. To validate the said model (FRM), landslide density index (LDI) has been computed.

14.2 MATERIALS AND METHODS

According to Putri et al. (2013) [27], disaster detection and monitoring are the important steps in disaster management and mitigation. The microwave remote sensing satellite imageries are also capable in mitigating the disaster as it can monitor hazard-prone areas regardless of the cloud cover [27].

Here, the landslide has been considered as the dependent factor while slope, elevation, NDVI and rainfall are considered as conditioning factors. The frequency ratio is calculated using the following formula:

$$Frequency\ ratio\ (FR) = \frac{Percentage\ of\ landslide\ pixels\ in\ a\ category}{Percentage\ of\ total\ pixels\ in\ the\ same\ category.} \quad (14.1)$$

Then, the relative frequency is calculated using the following formula:

$$Relative\ frequency\,(RF) = \frac{Factor\ \ Class\,(FR)}{\Sigma Factor\ \ Class\,(FR)} \quad (14.2)$$

In order to decipher the mutual interdependency among the independent or conditioning factors, the prediction rate (PR) is calculated using the following formula:

$$Prediction\ Rate = \frac{(Maximum\ RF - Minimum\ RF)}{Min\ (Maximum\ RF - Minimum\ RF)} \quad (14.3)$$

The values of prediction rate are used as weightages for various conditioning factor while preparing the landslide susceptibility model. The model is validated through the computation of LDI by using the following formula.

$$Landslide\ Density\ Index\ (LDI)$$
$$= \frac{Percentage\ of\ landslide\ pixels\ in\ a\ category}{Percentage\ of\ class\ pixels\ in\ each\ class\ on\ the\ susceptibility\ map.} \quad (14.4)$$

14.3 RESULTS AND DISCUSSIONS

14.3.1 Analysis of General Physical Parameters of the Study Area

Himalayan mountain systems are marked by heterogeneous geology, geomorphology, climate, soil, vegetation, land cover, and land use thus accelerating the rate of landslides.

14.3.2 Elevation

The study area is located in a rugged terrain. Here the maximum elevation is found to be 5,658 metres while the minimum elevation is 567 metres (Figure 14.2).

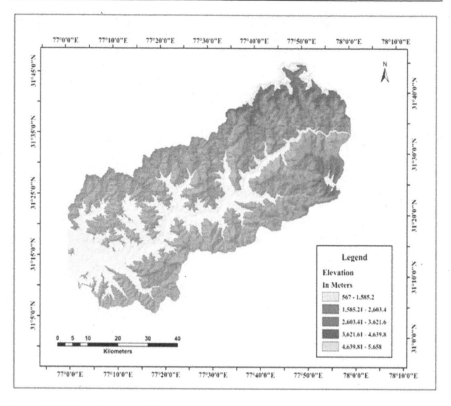

Figure 14.2 Elevation map of the study area.

14.3.3 Geology

Susceptibility of landslide is also portrayed by the geology. Dominance of Jeori-Wangtu banded gneissic complex is seen in the north-eastern and eastern part of the area under study and accounts for an area of 653.79 sq. km. Jutogh formations dominate the south-eastern and southern periphery of the study area accounting for a total area of about 242.01 sq. km (Figure 14.3).

14.3.4 Geomorphology

The study area is mainly dominated by highly dissected hills and valleys. Dominance of moderately dissected hills and valleys can be seen in the central part of the study area mainly around the river Sutlej. Deposition of alluvium is predominant mainly in the south-western part of the study area on either side of the Sutlej and its tributaries (Figure 14.4).

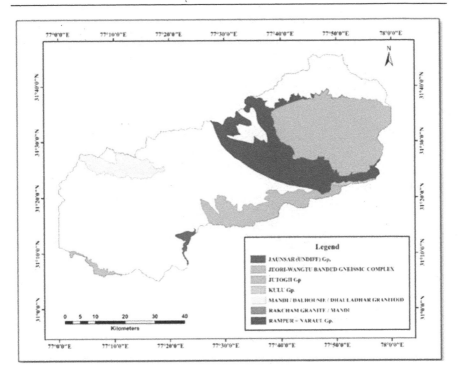

Figure 14.3 Geology map of the study area.

14.3.5 Slope

The study area is located in the region of varied slope. The value of slope ranges from the lowest value of 0° to the highest value of 77.57°. Most of the area under study is located within the slope 15.52°–31.03° and accounts for a total area of about 1600.4 sq. km (Figure 14.5).

14.3.6 Analysis of Drainage and Morphometric Properties of the Study Area

River Sutlej is the main river of the study area. There are mainly five order streams. The number of first-order streams is the highest of about 492 followed by 41 second-order streams, 17 third-order streams, 12 fourth-order streams, and 8 fifth-order streams (Table 14.1 and Figure 14.6).

Thus, the total number of streams present within the basin is 570. The total length of all the streams is found to be 1191.93 km (Table 14.2).

For a much better understanding of the nature and properties of the part of the drainage basin under investigation, morphometric analysis of the drainage basin is performed. The total area of the study area is found

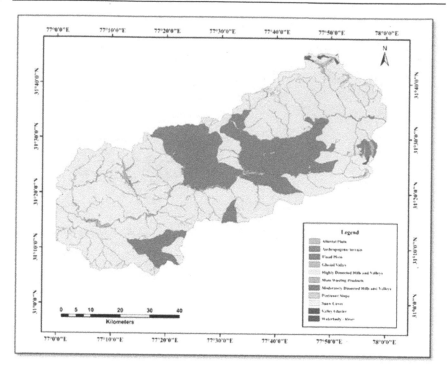

Figure 14.4 Geomorphology map of the study area.

to be 3671 sq. km. (Table 14.3). The perimeter of the basin is 345.1 km while the basin length is 105.40 km. The length of the main channel is identified as 51.16 km (Table 14.3).

14.3.7 Analysis of Land Use and Land Cover of the Study Area

Landslide susceptibility is highly determined by the land use and land cover scenario. The study area is dominated by trees (Figure 14.7).

14.3.8 Analysis of Rainfall Scenario of the Study Area

The amount of rainfall exhibits an increasing trend from the north-eastern to the south-western part of the basin (Figure 14.8).

14.3.9 Analysis of NDVI of the Study Area

NDVI is a statistical algorithm used to determine the vegetation health. Vegetation also takes a pivotal role in determining the degree of landslide susceptibility. The roots of vegetation bind the soil tightly with

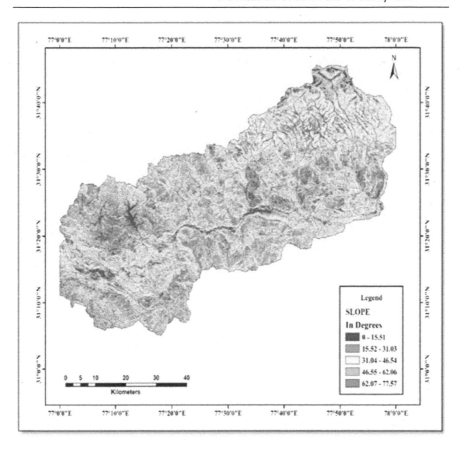

Figure 14.5 Slope map of the study area.

Table 14.1 Number of stream segments

Stream order	Number of stream segments
1	492
2	41
3	17
4	12
5	8
Total	**570**

the help of their roots and thus prevents the downslope movement in the hilly areas. In contrary, barren lands that are devoid of vegetation, the soil and loose materials become more exposed to various agents of erosion, become loose and fall down the slope under the gravitational

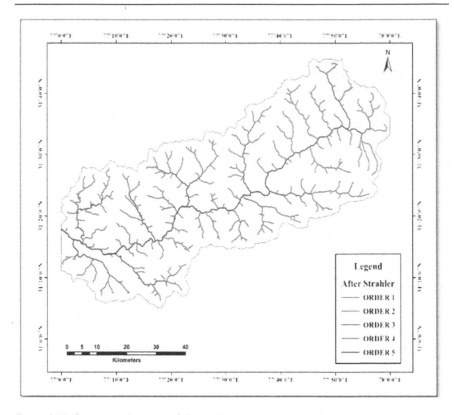

Figure 14.6 Stream order map of the study area.

Table 14.2 Stream length calculation

Stream order	Total length of streams (km)	Mean length of streams (km)
1	560.09	1.138394309
2	313.51	7.646585366
3	177.29	10.42882353
4	89.4	7.45
5	51.64	6.46
Total	**1191.93**	

influence and trigger landslides. In order to determine the degree of
landslides, it is very essential to take into consideration the vegetation
health trend. If the NDVI value ranges from 0.3 to 0.6, the vegetation is
considered to be stressed while if the NDVI value is more than 0.6, the
vegetation is considered to be healthy. The highest NDVI value of the

Table 14.3 Morphometric properties of the study area

Catchment size properties	Basin area (sq. km)	3,671
	Basin perimeter (km)	345.1
	Basin length (km)	105.4
	Main channel length (km)	51.16
	Stream order	5
Relief properties	Absolute relief (m) maximum	5,658
	Absolute relief (m) minimum	567
	Ruggedness number	1.65
	Relative relief (m)	5,091
	Relief ratio	0.04
Shape properties	Circularity ratio	0.39
	Elongation ratio	0.65
Texture properties	Drainage density (km/sq. km)	0.32
	Constant of channel maintenance (sq. km/km)	3.08
	Stream frequency (per sq. km)	0.16
	Bifurcation ratio	4.33

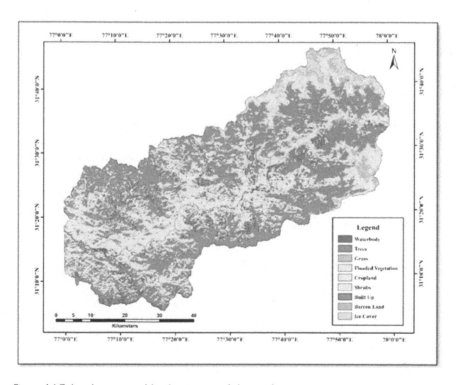

Figure 14.7 Land cover and land use map of the study area.

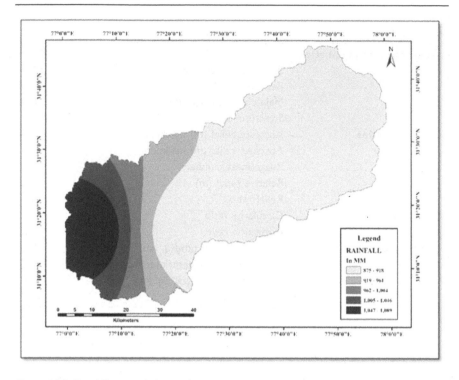

Figure 14.8 Rainfall map of the study area.

study area indicates that the vegetation status is healthy. However, on further classification, an interesting fact is noticed. The area under healthy vegetation (NDVI values of more than 0.6) is very less than about 0.0081 sq. km. Most of the vegetation within the study area is highly stressed (NDVI values ranging between 0.3 and 0.4) and covers an area of about 219.63 sq. km. The area under moderately stressed vegetation (NDVI values ranging from 0.4 to 0.5) and less stressed vegetation (NDVI value varying from 0.5 to 0.6) are 4.38 and 0.08 sq. km, respectively (Figure 14.9).

14.3.10 Appraisal of Landslide Susceptibility of a Part of Sutlej Basin

Landslide is one of the most devastating natural hazards that brings about colossal loss to human lives and property. In simpler words, landslide is defined as the downslope movement of weathered debris downslope under the impact of gravity. On the basis of the nature of material involved in landslide, they can be categorized into several manifestations such as fall, slides, topples, flows, and lateral spreads. Landslide is affected by a number of factors. However, here four major factors namely slope, elevation,

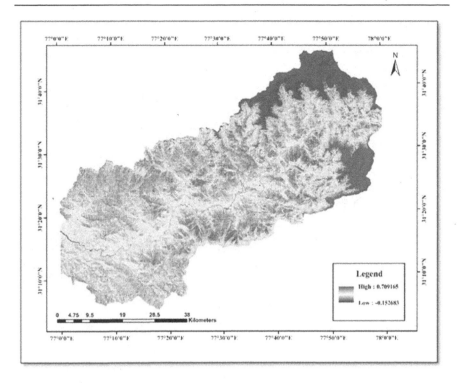

Figure 14.9 NDVI map of the study area.

vegetation health, and rainfall have been considered for the preparation of LSM (Table 14.4 and Figure 14.10).

14.3.11 Validating the Model

The analysis of LSMis validated using the LDI which is defined as the ratio of landslide pixels percentage to the class pixels percentage in each class of the LSM (Table 14.5).

Through the computation of LDI, it is seen that the highest value LDI is 5.52 and it falls within the moderate susceptibility zone having the landslide susceptibility index (LSI) values ranging between 212.11 and 343.91.

14.3.12 Role of Vegetation to PreventLandslides

Vegetation settles forested slants by giving root strength and by changing the soaked soil water system. There is two principal beneficial outcomes of the vegetation on slope stability; (14.1) Geo-mechanical effect, i.e., the effect of vegetation on incline soundness is to settle the slant with mechanical support of soil through roots. (14.2) Soil-hydrological effect, i.e., the plant

Table 14.4 Frequency ratio model

Class	Class pixel	Total pixels	Percent	Landslide pixel	Total pixels	Percent	FR	RF	RF (Non %)	RF (Int)
Slope										
1	411073	1022232	40.21327839	39	1650	2.36363636	0.05877751	0.009367823	—	1
2	177822		17.39546404	471		28.54545455	1.640971145	0.261534184	26.1534184	26
3	162620		15.90832609	907		54.96969697	3.455404211	0.550714327	55.07143273	55
4	263336		25.76088403	226		13.6969697	0.531696416	0.084740544	8.474054436	8
5	7381		0.722047441	7		0.42424242	0.587554778	0.093643121	9.364312096	9
							6.274404059	1		
Elevation										
1	986091	2366696	41.66530049	543	1650	32.90909091	0.789844079	0.115541064	11.55410644	12
2	191058		8.072773183	556		33.6969697	4.174150436	0.610608849	61.06088489	61
3	765901		32.36161298	349		21.15151515	0.653598916	0.095610661	9.561066081	10
4	321983		13.60474687	169		10.24242424	0.752856657	0.110130419	11.0130419	11
5	101663		4.295566477	33		2	0.465596333	0.068109007	6.810900693	7
							6.83604642	1		
NDVI										
1	564290	942474	59.87326971	358	1650	21.6969697	0.362381574	0.023070927	2.307092718	2
2	151476		16.07216751	874		52.96969697	3.295740723	0.209822463	20.98224627	20
3	197785		20.98572481	413		25.03030303	1.192729975	0.075934839	7.593483888	7
4	28870		3.063214476	4		0.242424242	0.079140473	0.005038457	0.503845731	1
5	53		0.005623497	1		0.060606061	10.77728988	0.686133314	68.61333139	68
							15.70728263	1		
Rainfall										
1	2792905	4083795	68.38994122	1091	1650	66.12121212	0.966826568	0.180558564	18.05585643	18
2	402725		9.861538104	161		9.757575758	0.989457796	0.184785033	18.47850329	19
3	224199		5.489967053	136		8.24242424	1.501361331	0.280384979	28.0384787	4
4	269539		6.600208874	87		5.272727273	0.798872789	0.149192553	14.912553	14
5	394427		9.6583475	175		10.60606061	1.098124045	0.205078871	20.50788711	20
							5.354642529	1		

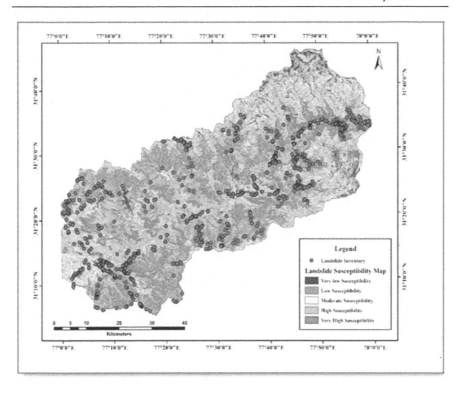

Figure 14.10 Landslide susceptibility map of the study area.

on slope stability is to reduce soil water content via interception, transpiration and evapotranspiration.

Shear stress (τ) can be expressed along a basal zone of sliding as:

$$\tau = W \sin \alpha \tag{14.5}$$

where W is the effective weight of the soil, and α is the slope of the failure surface.

Shear strength (S) can be expressed as:

$$S = C + W \cos \alpha \tan \varnothing \tag{14.6}$$

where C is the soil cohesion effectiveness and $\varnothing$ is the internal friction angle.

14.4 CONCLUSION

Disasters can occur at anytime and anywhere. Availability of proper forecast and preventive measures enable the administrative authorities to

Table 14.5 Computation of landslide density index

Class	LSI	Class pixels	% pixels	Landslide pixels	%landslide	Landslide density
Very low	59.86–141.66	217474	11.22944314	89	5.393939394	0.480338991
Low	141.66–212.11	595707	30.75980525	225	13.63636364	0.443317619
Moderate	212.11–343.91	107829	5.567836269	507	30.72727273	5.518709826
High	343.91–466.62	129239	6.673358666	469	28.42424242	4.259360818
Very high	466.62–639.32	886392	45.76955667	360	21.81818182	0.476696376
		1936641		1650	100	

take proper steps to save maximum damage and death tolls. LSM of this Sutlej river basin prepared using FRM clearly exhibits that the central, eastern and western part of the study area are very highly susceptible to landslides while the northern and north-eastern part is not so susceptible to landslides. The model also reveals that landslide in the basinal area is highly affected by the conditioning factors taken into account. All types of developmental activities such as urbanization and road construction have to be done keeping in mind the susceptibility to landslides in the area. The government has already taken those steps and it is clearly exhibited from the LDI value where it is seen that the area of very high susceptibility is having low LDI value of 0.47. More scientific and efficient planning strategies are required for undertaking developmental activities especially in moderate and highly susceptible areas with high LDI values of 5.52 and 4.26, respectively. Environmental impact assessment (EIA) has to be done before undertaking any big developmental project so that disastrous landslides in high and very high susceptible areas can be minimized and loss of life and property can be averted. EIA will enable in adopting proper disaster management strategies and prevent the occurrence of maximum damages. If all developmental activities are done keeping in mind the landslide susceptibility scenario as well as the EIA, it will enhance the development of tourism industry of the state especially eco-tourism. This will not only trigger the development of the study area but will also act as an indispensable tool for the future progress of not only the state of Himachal Pradesh but also India as a whole.

REFERENCES

[1] Rai, P. K., Mohanand, K., & Kumra, V. K. (2014). Landslide hazard and its mapping using remote sensing and GIS. *Journal of Scientific Research*, *58*, 01–13.

[2] Oliva-González, A. O., Pozo Ruiz, A. F., Amaya Gallardo, R. J., & Jaramillo, H. Y. (2019). Landslide risk assessment in slopes and hillsides. Methodology and application in a real case. *DYNA*, *86*(208), 143–152 (accessed 10 March 2022) [Online]. 10.15446/dyna.v86n208.72341

[3] De, S. K. (2004). Causes of landslide in the balason basin of eastern himalaya. In Singh, S., Sharma, H. S., & De, S. K. (eds.), *Geomorphology and environment* (pp. 182–194). Kolkata: ACB Publications.

[4] Biswas, S. S., & Pal, R. (2016). Causes of landslides in Darjeeling Himalayas during June-July, 2015. *Journal of Geography & Natural Disasters*, *6*(2), 01–05 (accessed 10 March 2022) [Online]. 10.4172/2167-0587.1000173

[5] Chawla, A., Chawla, S., Pasupuleti, S, Rao, A. C. S., Sarkarand, K., & Dwivedi, R (2018). Landslide susceptibility mapping in Darjeeling Himalayas, India. *Advances in Civil Engineering*, *2018*, 01–17 (accessed 10 March 2022) [Online]. 10.1155/2018/6416492

[6] Gerrard, J. (1994). The landslide hazard in the Himalayas: geological control and human action. *Geomorphology, 10*, 221–230.

[7] Khan, R., Yousaf, S., Haseeb, A., & Uddin, M. I. (2021). Exploring a design of landslide monitoring system. *Complexity, 2021*, 01–13 (accessed 10 March 2022) [Online]. 10.1155/2021/5552417

[8] Yang, M., Xiong, S., Cao, Y., Ling, Z., Liu, Y., & Xie, Y. (2020). Landslide hazard evaluation based on linear rupture plane method. *E3S Web of Conferences, 143*, 01–04 (accessed 10 March 2022) [Online]. 10.1051/e3 sconf/20201430

[9] Pardeshi, S. D., Autade, S. E., & Pardeshi, S. S. (2013). Landslide hazard assessment: recent trends andtechniques. *Springer Plus, 2*(523), 01–11.

[10] Dikshit, A., Sarkar, R., Pradhan, B., Segoni, S., & Alamri, A. M. (2020). Rainfall induced landslide studies in Indian-Himalayan region: a critical review. *Applied Sciences, 10*(2466), 01–24 (accessed 11 March 2022) [Online]. 10.3390/app10072466

[11] Schuster, R. (1996). Socioeconomic significance of landslides. In Turner, A. K. & Schuster, R. L. (eds.), *Landslides: Investigation and Mitigation: Special Report* (*vol. 247*, pp. 12–36). Washington, DC, USA: National Academic Press.

[12] Chae, B. G., Park Jin, H., Catani, F., Simoni, A., & Berti, M. (2017). Landslide prediction, monitoring and early warning: a concise review of state-of-the-art. *Geosciences Journal, 21*(6), 1033–1070 (accessed 10 March 2022) [Online]. 10.1007/s12303-017-0034-4

[13] Arbanas, Z., Sassa, K., Nagai, O., Jagodnik, V., Vivoda, M., Jovancevic, S. D., Peranic, J., & Ljutic, K. (2014). A landslide monitoring and early warning system using integration of GPS, TPS and conventionalgeotechnical monitoring methods. In *Proceedings of World Landslide Forum*, Beijing, 01–06.

[14] Vakhshoori, V., & Zare, M. (2016). Landslide susceptibility mapping by comparing weight of evidence, fuzzy logic, and frequency ratio methods. *Geomatics, Natural Hazards and Risk, 7*(5), 1731–1752.

[15] Matternicht, G., Hurni, L., & Gogu, R. (2005). Remote sensing of landslides: an analysis of the potential contribution to geospatial systems for hazard assessment in mountainous environments. *Remote Sensing of Environment, 124*, 348–359.

[16] Zhao, C., & Lu, Z. (2018). Remote sensing of landslides—a review. *Remote Sensing, 10*(279), 01–06 (accessed 10 March 2022) [Online]. 10.3390/rs1 0020279

[17] Brabb, E. (1993). Proposal for worldwide landslide hazard maps. In *Proceedings of 7th International Conference and field workshop on landslide in Czech and Slovak Republics*, 15–27.

[18] Fayez, L., Pham, B. T., Solanki, H. A., Pazhman, D., Dholakia, M. B., Khalid, M., & Prakash, I. (2018). Application of frequency ratio model for the development of landslide susceptibility mapping at part of Uttarakhand State, India. *International Journal of Applied Engineering Research, 13*(9), 6846–6854.

[19] Bin, L., Yu, B., Chen, F., Shakir, M., Wang, L., & Wu, M. (2017). A simple but effective landslide detection method based on image saliency. *Photogrammetric Engineering and Remote Sensing, 83*(5), 351–363.

[20] Elmoulat, M., & Brahim, L. A. (2018). Landslides susceptibility mapping using GIS and weights of evidence model in Tetouan-Ras-Mazari area (Northern Morocco). *Geomatics, Natural Hazards and Risk*, *9*(1), 1306–1325. (accessed 10 March 2022) [Online]. 10.1080/19475705.2018. 1505666

[21] Vojtekova, J., & Vojtek, M. (2020). Assessment of landslide susceptibility at a local spatial scale applying the multi-criteria analysis and GIS: a case study from Slovakia. *Geomatics, Natural Hazards and Risk*, *11*(1), 131–148 (accessed 10 March 2022) [Online]. 10.1080/19475705.2020.1713233

[22] Berhane, G., Kebede, M., & Alfarrah, N. (2020). Landslide susceptibility mapping and rock slope stability assessment using frequency ratio and kinematic analysis in the mountains of Mgulat area, Northern Ethiopia. *Bulletin of Engineering Geology and the Environment*, *80*(3), 1–17 (accessed 10 March 2022) [Online]. 10.1007/s10064-020-01905-9

[23] Nohani, E., Moharrami, M., Sharafi, S., Khosravi, K., Pradhan, B., Pham, B. T., Lee, S., & Melesse, A. M. (2019). Landslide susceptibility mapping using different GIS-based bivariate models. *Water*, *11*(1402), 01–22 (accessed 10 March 2022) [Online]. 10.3390/w11071402

[24] Igwe, O., & Ozioko, O. (2020). GIS-based landslide susceptibility mapping using heuristic and bivariate statistical methods for Iva Valley and environs Southeast Nigeria. *Environmental Monitoring and Assessment*, *192*(2) (accessed 10 March 2022) [Online]. 10.1007/s10661-019-7951-9

[25] Anis, Z., Wissem, G., Riheb, H., Vali,. V., Smida, H., & Essghaier, G. M. (2019). GIS-based landslide susceptibility mapping usingbivariate statistical methods in North-western Tunisia. *Open Geosciences*, *11*(1), 708–726 (accessed 10 March 2022) [Online]. 10.1515/geo-2019-0056

[26] Ahmad, B. B., Khezri, S., Shahabi, H., & Baharin, B. (2013). Landslide susceptibility mapping in central Zab basin in GIS-based models, Northwest of Iran. *Journal of Basic and Applied Scientific Research*, *3*(3), 924–930.

[27] Putri, R. F., Wibirama, S., Allimuddin, I., Kuze, H., & Sumantyo, J. T. S. (2014). Monitoring and analysis of landslide hazard using Dinsar technique applied to AlosPalsar Imagery: a case study in Kayangan Catchment Area, Yogyakarta, Indonesia. *Journal of Urban and Environmental Engineering*, *7*(2), 308–322 (accessed 12 March 2022) [Online]. 10.4090/juee.2013.v3n2. 308322

Chapter 15

Crop Recommendation System Using Machine Learning Algorithms

Pranjali G. Gulhane and Anuradha Thakare

CONTENTS

15.1 INTRODUCTION

In India, farming is the primary occupation of many families. Almost half of India's manpower has an occupation which is directly or indirectly dependent upon agriculture. That is why India is rightfully called as an agricultural country. As per government data agriculture sector contributes near 20% contributions to the Gross Domestic Product of the country [1,2].

DOI: 10.1201/9781003355960-15

India is among the world's biggest producers of agricultural production. Though India is a leading agricultural country, its productivity per acre is very less as compared to China and other Asian countries. Farmer's income is very less due to the lack of farm productivity. It has been noticed that a farmer's crop decision is typically driven by his intuition and irrelevant considerations such as pricing in the previous season, lack of understanding about the market mandate, overestimation of a soil's capacity to sustain a specific crop, and so on [3]. The decision generally goes wrong leading to the worst financial condition of the farmer's family. This is one of the major reasons for farmer suicides happening in the country. Therefore, it is necessary to design a system that can recommend the crops which will give more yield in the season. This can be achieved by incorporating the latest technology into the design of the recommendation system.

Precision agriculture is defined as identifying parameters that are site-specific to resolve the issues in the selection of crops. The "site-specific" technique definitely improves the results, but supervision of such results is necessary for the system. It is not possible to provide accurate results by means of precision agriculture. However, in agriculture, exact and precise advice is required, as mistakes can result in material and monetary losses for farmers. Much efforts have been made in order to build a model that can properly and effectively propose crops. These research works used ensembling as one of the techniques for developing the system. Thus, the need is highlighted to combine the latest technologies and methodologies to create a new model to help the real issues of farmers.

For the various machine learning techniques used for development, we have proposed the system using neural networks (NN). Using XGBoost algorithm, we propose the system here which is efficient and accurate to recommend the crops to farmers.

15.1.1 Motivation

Agriculture is a significant part of the Indian economy. The excessive use of pesticides has reduced the fertility of soil considerably. Industrialization also reduced the area under agriculture. Many of the traditional methods used by agriculturists are not sufficient to raise productivity. The difficulty Indian farmers face is that, insufficient information about crop recommendations as per their soil and hence productivity hampers.

Indian farmers have insufficient data to decide on the selection of farming techniques and crops to adopt based on climate conditions and the state of the soil. The common problem observed in farmers is that the proper crops are not selected to get maximum production based on topographic features and financial viability. To get maximum crop production with minimum investment in inputs and processing are challenges in the agriculture sector.

15.1.2 Problem Statement

Farmers' inability to select proper crops using traditional and non-scientific methods for their land leads to serious problems in our country, where a large portion of the population is dependent on farming. The correct information is not available and accessible to researchers in agriculture to develop proper case studies. With resource constraints, we presented a method that tackles this issue by delivering predictive insights about the crop. Machine learning algorithms educated on critical environmental and economic characteristics are used to propose sustainable crops. Our aim is to develop a recommendation system for crops based on big data analytics by considering the factors like weather, water availability, geographical features and soil type. To provide a simpler technique for predicting which crops are suited for growing in a certain soil [4].

15.1.3 Objectives

The main objectives of the article are as follows:

- To create a crop prediction model based on soil and environmental data.
- By using ensemble techniques to improve the yield of crops by predicting with accuracy and efficiency.
- Apply precision agricultural concepts to eliminate crop selection errors.
- Create a comprehensive model capable of accurately predicting crop sustainability in a given soil and meteorological condition.
- Providing recommendations for appropriate crops in the vicinity so that the farmer suffers no financial losses.
- To provide analysis of different crops using data of previous years.

To help the farmers to resolve the queries such as financial help and crop selection, recommend them different government schemes so that farmers will get help and know the process to get eligible for the scheme.

15.2 RELATED WORK

Various research has been conducted in the recent past to build and implement the crop recommendation system. Each task that has been implemented and published in the past has its own set of advantages and downsides. The systems were created with the specific challenge of crop improvement recommendations.

Many scholars have begun to recognize problems in the Indian agriculture industry and are devoting more time and effort to resolving them.

The paper [5] describes the research and development of an accurate agronomic production predicting method based on actual monthly meteorological data. Because of the unusual weather that occurs every year, as well as the speedy regional climate violation caused by global warming, agricultural crop output is impossible to anticipate. The development of a system for forecasting agricultural yields based on actual meteorological data. The authors of this study explain how to handle various weather data like monthly or daily and also discussed how the predictive model is designed. They develop a non-parametric prediction approach based on 33 years of agricultural meteorological data. They can anticipate final production using monthly meteorological data, according to the applied model.

The study [6] takes into account pan-India farmers in order to present an intelligent crop recommendation system that is simple to design and utilize. This system will take into account a number of environmental and geographical criteria to help farmers make pre-stored decisions on which crop to cultivate. They have also created a subsystem to anticipate rainfall, which delivers rainfall forecasts for the following 12 months. The study [7] provides a recommendation system based on a prediction classifier with a maximum rule that employs Random tree, CHAID, K-Nearest Neighbour, and Naive Bayes as learners to propose a crop for site-specific characteristics with significant efficiency and accuracy.

The study [8] used a neural network to forecast agricultural yield. The system is based on the idea of linear predictive models for analyzing yield and the best crop to plant. The model was trained to provide the optimum result using the training and testing data. The algorithm processed the data and generated a forecast profitable crop, a list of required fertilizers, and a complete overall yield per hectare. With an accuracy of 85%, the algorithm predicted solely the yield of the selected crop.

The paper [9] showcases the crop prediction system based on sensor networks based on IoT. The main issue in soil testing is the time required to get results of soil condition status. Hence, the author claims that the system will help the farmers to get quick crop prediction for selection without much delay or waiting. The projected method in the article sensor network collects data which is processed to get nutrients contents efficiently from the soil. This enables in selecting proper crop for the soil that is examined. The farmers enlist their IoT-based NPK sensor with the primary server. The NPK extracts level from the soil sample are stored and refreshed this information to the main primary server through the Raspberry Pi unit of sensor. The reading collected is processed and crops are predicted with the help of stored reading. The drawback of this method has limited sensor parameters hence some incompetence in crop prediction is observed in process and data collected through NPK-IoT sensor are not steady which give problems in value selection. A large emphasis is placed on data collecting using the NPK sensor, which has a wide range of variations.

In paper [10], the authors created a rain forecast using polynomial regression for the Farming Sector Forecast for Karnataka. Our end result is the best combination of rain and crop kind. Using solely rain forecast as input, the machine proposes crop. The suggested approach is only applicable to the Indian state of Karnataka.

Environmental factors such as humidity, temperatures, and state geographical region, as well as soil parameters such as type of soil, pH level, and levels of nutrients, are considered in the paper [11] to recommend an acceptable crop to the farmer. The system in [12] was created using ML and is one of the AI applications that allows systems to learn and adapt automatically without the need for human intervention. Following that, the program's accuracy will be increased deprived of the need for human interaction. For corps ranking, regression methods were utilized. The random forest method and BigML were used to analyze crop characteristics. ML algorithms were utilized to mitigate the effects of water stress in plants and provided a group of decision rules for plant state forecasting.

Existing systems that employ machine learning algorithms are beneficial and have high accuracy when applied in their respective locations. Existing methods mostly employ soil properties as input for crop prediction. These systems' primary goals were yield prediction rather than crop suggestion. The suggested system's input parameters include environmental conditions and soil characteristics to recommend the best crop for maximizing yield and enhancing agriculture productivity for rural farmers in India. Furthermore, the suggested mechanism for crop knowledge categorization to farmers. Furthermore, existing systems' accuracy is poor in comparison to our work. Compared to the suggested system, it provides more accuracy.

15.3 METHODOLOGY

To solve the aforementioned problems, we offer an Intelligent Crop Recommendation system that predicts crop appropriateness by taking into consideration all essential data such as temperature, rainfall, location, and soil condition. The whole system is intrinsically focused on executing the core purpose of the developed framework, which is to deliver crop suggestions to farmer's algorithms. We also give profit analyses on crops cultivated in various states, providing the customer with a simple and dependable insight into crop selection and planning. The proposed system offers a solution for selecting the most suitable crop based on temperature. The values are fed into the system as input, and the system analyses the data to produce crop results as output. The approach recommended that crops be cultivated effectively in order for farmers to optimize productivity and profit. Our approach also gives information about crop seasonal categorization, allowing us to decide which crops grow effectively and healthily in whatever environment and soil. Our method has an accuracy rate of 90% on average on the supplied dataset.

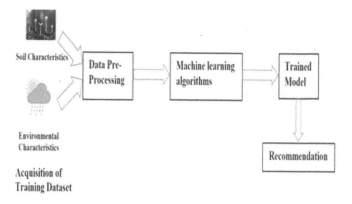

Figure 15.1 System architecture.

Our suggested research takes into account both environmental and soil characteristics. This is because a certain type of soil will sustain a crop but weather circumstances will not, resulting in a reduced yield. The complete operation of the suggested system is depicted in Figure 15.1.

15.3.1 Data Analysis

This is an endeavour to uncover any relationships between the dataset's various properties.

15.3.1.1 Acquirement of Training Dataset

The algorithm is fed data from the government website [13] and Kaggle. Among the datasets are the following: (i) yield dataset: it includes yield in kilograms per hectare for 16 primary crops cultivated in all states. A yield of zero means that the crop is not planted. (ii) Cost of cultivation dataset: this set of data covers the cost of cultivation in rupees per hectare for each crop. (iii) Crop modal price: this set of data includes the average price level for those crops for more than a two-month period. (iv) Crops standardized price: this information provides the market rate of crops in rupees per hectare. (v) Soil nutritive value data: this set of data has five columns, each with the following characteristics: state, nitrogen content, phosphorous content, potassium content, and mean pH within this order. This information comprises crops, min and max rainfall, min and max temperature, and pH values. The cultivation cost, market cost, standard cost, and yield dataset are used for profit analysis. This is done as a first step to see how big of an influence profit may have on crop forecast. The profit for each crop cultivated in the state is computed and states that produce zero or no crop are awarded a –1 value.

15.3.2 Data Processing

This phase entails changing the null and 0 yield values with −1 to make sure that the overall forecast remains unchanged. Moreover, before the dataset can be fed into the neural network, it must be encoded. Data pre-processing is critical since it cleans the data and prepares it for usage in ML algorithms. Pre-processing focuses mostly on eliminating outliers and erroneous data, as well as coping with missing information. The dataset's values are mostly in string format. This should be converted into integer values before being fed into the neural network. To restrict the amount of data feeding into the linear regression model, crops are screened based on the nutrients required and present in the soil. If the soil's nutrient level is lower than what the crops require, the crop will be abandoned; as a consequence, the training time has been substantially shortened.

15.3.3 Training Model and Crop Recommendation

During pre-processing, the set of data is used to train various ML models, such as neural networks and linear regression, to achieve the highest possible accuracy.

15.4 XBOOST ALGORITHM

XGBoost: It is a Tree Boosting System which is more scalable

XGBoost is a newly dominant technique in applied ML and Kaggle contests for organized or tabular data. XGBoost is a gradient-boosted decision tree implementation optimized for speed and performance.

The two main reasons to utilize XGBoost:

1. Execution speed
2. Model performance

1. XGBoost Execution Speed

In general, XGBoost is efficient. This approach is exceptionally fast when compared to the previous gradient boosting techniques.

2. XGBoost Model Performance

In classification and regression computational modelling issues, XGBoost surpasses organized or hierarchical information. As per the facts, it is the preferred method for contest winners on the competitive data science platform Kaggle.

15.4.1 Algorithm Demonstration

In the initial phase, the system must be configured with a function $F_0(x)$. $F_0(x)$ would be a function that optimizes the loss function or MSE (mean squared error) in this case:

Equation 1:

$$F_0(x) = \text{argmin}_\Gamma \sum_{i=1}^{n} L(Y_i, \Gamma)$$

Equation 2:

$$\text{argmin}_\Gamma \sum_{i=1}^{n} L(Y_i, \Gamma) = \text{argmin}_\Gamma \sum_{i=1}^{n} (Y_i - \Gamma)^2$$

The function diminishes at the mean $i=1nyin$ when the primary discrepancy of the overhead equation is taken with respect to γ. As a result, the boosting model could begin with:

Equation 3:

$$F_0(x) = \left(\sum_{i=1}^{n} Y_i \right) / n$$

15.5 EXPERIMENTAL AND PERFORMANCE MATRIX

15.5.1 Experimental Setup

Anaconda Individual Edition 2020.11 (Jupyter Notebook) IDE and Python 3.9.0 are used to run the experiment. Numpy, pandas, Sklearn.tree, Sklearn.ensemble>RandomForestClassifier, and Sklearn.ensemble>Gradient BoostingClassifier were also used to carry out the experiments. In terms of hardware configuration, we used an Intel i5 processor with 8GB of RAM.

15.5.2 Dataset

The dataset used in this study was obtained from Kaggle and government websites, which provide statistical data on soil contaminants and the surrounding environment. This dataset includes soil contaminants such as nitrogen, phosphorus, and potassium levels. Temperature, humidity, water pH level, and rainfall are also included in the dataset for the surrounding environmental measures which can be seen in Table 15.1.

Table 15.1 Dataset description

Nitrogen level (N)	Phosphorus level (P)	Potassium level (K)	Temperature	Humidity	pH	Rainfall	Label
90	42	43	20.87974371	82.00274423	6.502985292	202.9355362	Rice
85	58	41	21.77046169	80.31964408	7.038096361	226.6555374	Rice
71	54	16	22.61359953	63.69070564	5.749914421	87.75953857	Maize
61	44	17	26.10018422	71.57476937	6.931756558	102.2662445	Maize
80	43	16	23.55882094	71.59351368	6.657964753	66.71995467	Maize
73	58	21	19.97215954	57.68272924	6.596060648	60.65171481	Maize
61	38	20	18.47891261	62.69503871	5.970458434	65.43835393	Maize
2	51	17	25.86768261	45.96341933	5.838508699	38.53254678	Mothbeans
16	51	21	31.01963639	49.9767522	3.532008668	32.81296548	Mmothbeans
19	55	20	27.43329405	87.80507732	7.18530147	54.73367631	Mungbean
55	77	22	31.4345059	62.99303471	7.76061831	64.77651469	Blackgram
42	79	23	27.71678273	63.29103387	6.781841984	68.56507978	Blackgram
44	77	21	32.63918668	61.3009051	7.326980454	61.83876146	Blackgram
38	62	25	32.7477393	67.77954584	7.453975408	63.37784443	Blackgram
32	76	15	28.05153602	63.49802189	7.604110177	43.35795377	Lentil

15.6 EXPERIMENTAL RESULTS

15.6.1 Exploratory Data Analysis

We must deal with missing values during data analysis. Handling missing data is so important that it deserves its own book. However, before we can do anything about missing values, we must first understand the pattern of missing value occurrence which can be seen in Figure 15.2.

A symmetrical dataset has an equal or nearly equal number of samples from both the positive and negative classifications. If one class contains higher samples than another, the data are biased in favour of one of the classes depicted in Figure 15.3.

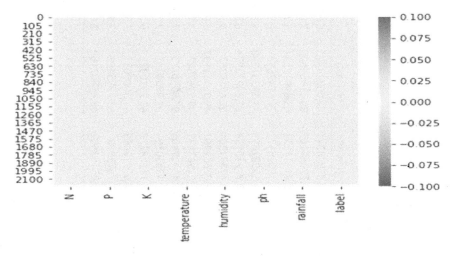

Figure 15.2 Heatmap to check null/missing values.

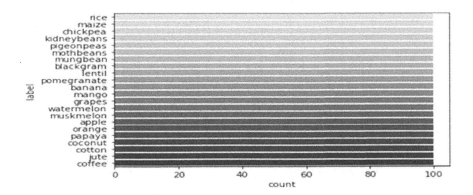

Figure 15.3 Dataset balance count plot.

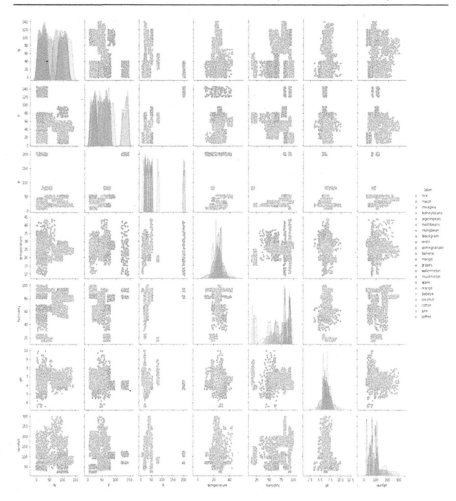

Figure 15.4 Diagonal distribution between two features.

The pairs plot shows both the distribution of single variables and the relationships between two variables. Variables are dataset features such as nitrogen level, rainfall, pH values, and so on, as shown in Figure 15.4.

In the margins, a joint plot depicts a relationship between multiple variables (bivariate) as well as 1D profiles (univariate). The variable here is the resultant column for all of the features as seen in Figure 15.5.

15.6.2 Classification Algorithm Accuracy Comparison

This dataset is being run with the models listed in Table 15.2. Each of these executions provided a different level of accuracy based on the algorithm's

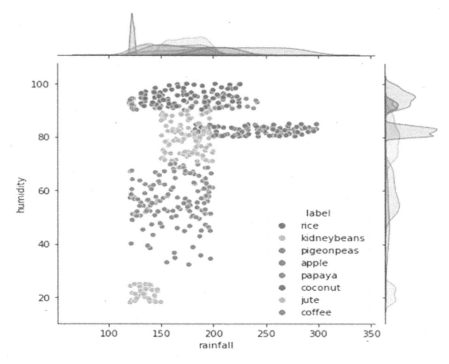

label
● rice
◉ kidneybeans
◉ pigeonpeas
● apple
● papaya
● coconut
◉ jute
● coffee

Figure 15.5 Joint plot.

Table 15.2 Classification of algorithms

Algorithm	Accuracy
KNN classifier	0.84
Support vector classifier	0.86
Decision tree	0.82
Random Forest	0.87
Xtreme gradient boosting	0.94

performance and execution timing. To obtain these results, all of them are classification algorithms that have been tested in the same environment. We discovered that the Xtream Gradient Boosting algorithm produced the best results in the shortest amount of time when compared to all other classification algorithms that can be seen in Table 15.2.

15.8 CONCLUSION

We successfully formulated and implemented an intelligent crop recommendation system for farmers across India in this chapter. India is a country where agriculture is very important. The prosperity of the farmers leads to the prosperity of the nation. As a result of our work, farmers will be able to sow the proper seed based on soil conditions, enhancing national output.

The suggested approach aids farmers in crop selection by offering insights that regular farmers do not keep track of, hence lowering crop failure and boosting production. It also protects them from financial loss. The model suggested in this research might be expanded in the future that forecasts crop rotations.

REFERENCES

[1] Kumar, R., Singh, M. P., Kumar, P., & Singh, J. P. (2015). Crop selection method to maximize crop yield rate using machine learning technique. In *2015 International Conference on Smart Technologies and Management for Computing, Communication, Controls, Energy, and Materials (ICSTM)*, Chennai (pp. 138–145). 10.1109/ICSTM.2015.7225403

[2] Lekhaa, T. R. (2016). Efficient crop yield and pesticide prediction for improving agricultural economy using data mining techniques. *International Journal of Modern Trends in Engineering and Science (IJMTES)*, 03(10), 11–28.

[3] Rajeswari, S. R., Khunteta, P., Kumar, S., Singh, A. R., & Pandey, V. (2019). Smart farming predict ion using machine learning. *International Journal of Innovative Technology and Exploring Engineering*, 08(07), 250–270.

[4] Gholap, J., Ingole, A., Gohil, J., Gargade, S., & Attar, V. (2012). Soil data analysis using classificat ion techniques and soil at tribute prediction. *International Journal of ComputerScience Issues*, 9(3), 560–578.

[5] Lee, H., & Moon, A. (2014). Development of yield prediction system based on real-time agricultural meteorological information. In *16th International Conference on Advanced Communication Technology*, Pyeongchang (pp. 1292–1295). 10.1109/ICACT.2014.6779168

[6] Doshi, Z., Nadkarni, S., Agrawal, R., & Shah, N. (2018). Agro consultant: intelligent crop recommendation system using machine learning algorithms. In *Fourth International Conference on Computing Communication Control and Automation (ICCUBEA)*, Pune, India (pp. 1–6). 10.1109/ICCUBEA. 2018.8697349

[7] Pudumalar, S., Ramanujam, E., Rajashree, R. H., Kavya, C., Kiruthika, T., & Nisha, J. (2017). Crop recommendation system for precision agriculture. In *Eighth International Conference on Advanced Computing (ICoAC)* (pp. 32–36). 10.1109/ICoAC.2017.7951740

[8] Thomas, K. T., Varsha, S., Saji, M. M., Varghese, L., & Thomas, E. J. (2020). Crop prediction using machine learning. *International Journal of Future Generation Communication and Networking*, 13(3), 1896–1901.

[9] Kumar, R., & V., Singhal (2022). IoT enabled crop prediction and irrigation automation system using machine learning. *Recent Advances in Computer Science and Communications (Formerly: Recent Patents on Computer Science)*, 15(1), 88–97.

[10] Liakos, K. G., Busato, P., Moshou, D., Pearson, S., & Bocht, D. (2018). Machine learning in agriculture: a review. *Article on Sensors*, 1–29. 10.3390/s18082674

[11] Shakoor, M. T., Rahman, K., Rayta, S. N., & Chakrabarty, A. (2017). Agricultural production output prediction using supervised machine learning techniques. In *1st International Conference on Next Generation Computing Applications (NextComp)*, Mauritius (pp. 182–187). 10.1109/NEXTCOMP.2017.8016196

[12] Bondre, D., & Mahagaonkar, S. (2008). Prediction of crop yield and fertilizer recommendation using machine learning algorithms. *International Journal of Engineering Applied Sciences and Technology*, 371–376. 10.33564/IJEAST.2019.v04i05.055.

[13] Savvas, D., & Goumopoulos, C. (2008). Applying machine learning to extract new knowledge in precision agriculture applications. In *2008 Panhellenic Conference on Informatics* (pp. 100–104). IEEE.

Chapter 16

Impacts of Artificial Intelligence, Deep Learning, and Internet of Things

Dr. Renu Bala

CONTENTS

16.1 INTRODUCTION

Under artificial intelligence (AI), machines perform activities similar to human characteristics. The designing of such kinds of machines requires hard efforts such as proper planning, problem-solving, identification and reorganization of sounds and objects, understanding language, etc. AI has taken place in each and every aspect of life, especially in the field of internet technology and businesses field [1]. Deep learning (DL) is considered as various approaches to machine learning. The term deep in DL represents various layers of a network that helps in complex processing. As DL generates an artificial neural network which involves various type of layers between input and output. In general, it is understood that DL is a subset of machine learning and both deep and machine learning come under the umbrella of AI as shown in Figure 16.1. The figure clears that AI is a broad concept and it further includes machine learning and DL as its components. AI concept is used for the creation of smart machines, machine learning helps to generate AI-driven applications; whereas DL uses a big amount of data as well as complex algorithms for the training of that model which has been implemented by AI. This is the main reason both machine learning and DL are considered as components of AI.

DOI: 10.1201/9781003355960-16

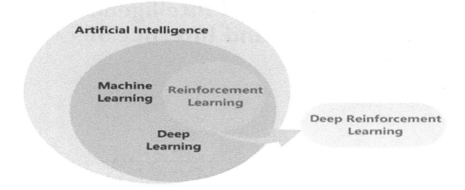

Figure 16.1 The relationship between artificial intelligence, machine learning, and deep learning.

But both the terms are different in various areas such as algorithms, complexity, human interference, scalability and human interference, etc. [a]. The Internet of Things (IoT) represents a system of interconnected digital machines, computing devices, people or animals and other objects. etc., which have the capabilities of transferring data over the network without requiring the interaction of human-to-computer or human-to-human. It can be understood in Figure 16.2. The figure represents that under the IoT technology how living and non-living things like machines, etc., connect with each other. As IoT provides the connection facility between man to machine, machine to machine, or anything to everything. In simple words, all devices are connected to the world with the help of IoT services.

Sensors, antenna, microcontrollers, etc., are good examples of IoT devices; under an IoT system, such kinds of devices are used to collect data. The IoT offers chances to collect continuous data regarding every physical activity of the business [3]. After the collection of data, it is collated as well as transferred by the IoT hub and gateway and at the end analyzed by

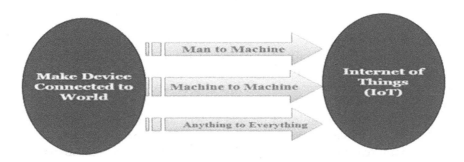

Figure 16.2 Nature of IoT [2].

human machines, smartphones, consumer relationship management, etc. [b]. For the prediction of future natural hazards, IoT is considered as the best effective tool after combining it with DL [4]. In general, IoT has reformed every field of life after the emergence of AI [5]. The concept of IoT has completely transformed the old style of living into a high-tech standard [6].

16.2 LITERATURE REVIEW

Madakam et al. [7] explained the uses of IoT on the basis of previous research and found that such technology is gaining popularity day by day. Osuwa et al. [8] explained the applications of AI in the field of IoT. The results of the study revealed that AI is used in the field of IoT in connected ways to fulfil various applications. Sarmah et al. [9] elaborated the meaning of the term IoT with suitable examples as well as highlighted the advantages of IoT. The study showed that in the present era, IoT has become a necessity in various fields. Ahmed [10] defined the advantages and disadvantages of AI. The analyzed data presented that besides a lot of advantages; AI has some disadvantages also. Boutaba et al. [11] presented the relationship between DL and machine learning and research opportunities of machine learning for future time period. The outcomes of the research indicated that DL and machine learning both have bright future due to their implications. Ghosh et al. [12] explained how AI and IoT both are interrelated with each other. After the analysis of collected data, it has been observed that both kinds of technologies are interrelated with each other. Olly [13] described the trend of AI in the United States, Europe, and China. The outcomes of the study indicated that in these countries the popularity of AI is increasing day by day due to its vast applications. Hussein [14] defined the future applications of IoT such as water management, smart agriculture, healthcare, retail and smart living, etc. Kankanhalli et al. [15] highlighted the benefits of machine learning and AI in the field of marketing on the basis of various packaging and distribution abilities, etc. Kankanhalli et al. [15] elaborated on the smart uses of AI and IoT in the government services sector. The analyzed data reported that in government agencies the uses of such technologies are increasing day by day, but it has some drawbacks also as it is reducing the manpower requirement. Kumar et al. [16] highlighted the advantages of IoT and considered it as revolution for the future time periods. Pal [17] described the application of IoT specifically in healthcare sector. It has been revealed that in this sector IoT is providing a lot of benefits and in the future time period, the demand for IoT will increase for healthcare services. Sharma and Nandal [18] described the fusion of machine learning and IoT. The facts described that by the fusion of machine learning and

IoT various advantages can be taken from the latest technology. Brous et al. [19] explained the limitations of IoT in the form of security issues for big organizations. Goyal et al. [20] elaborated that how machine learning is implemented in IoT. It has been found that machine learning and IoT are related to each other in different aspects. Merenda et al. [21] described the importance of edge machine learning regarding AI-enabled devices of IoT. The analyzed data presented that in the future time period almost all devices will be designed on the basis of machine learning to take accurate decisions and to make predictions. Mufti et al. [22] explained the different kinds of challenges and applications of IoT from the cloud perspective. The study found that there are a lot of areas where such kinds of technologies are in great demand and the future of these technologies is very bright. Bala [23] showed the perspectives of teachers and students of different higher educational institutions re-garding cloud-based digitalization. The result of the study suggested that cloud-based digitalization has both positive and negative aspects; so the focus should be made towards the positive aspects by the avoidance of negative side. Bolhasani et al. [24] highlighted the role of IoT specifically in healthcare services. During the study, it has been founded that the use of IoT technology is increasing in the healthcare services as different kinds of reputed organizations in such areas are promoting IoT for their services. Kuzlu et al. [25] described the role of AI in the IoT cyber se-curity. It has been founded that AI has a leading role in such an area. Sultana and Tamanna [26] analyzed the benefits and challenges of IoT in the situation of COVID-19. In the study, it has been found that IoT provided help to maintain social distancing by avoiding personal communication. Toorajipour et al. [27] elaborated the uses of AI and IoT in supply chain management (SCM) on the basis of old related studies. The outcomes indicated that the network-based structure of SCM provides a logical framework for the implementation of AI. Zuiderwijk et al. [28] showed the implications of the use of AI in places related to public governance. The analyzed data revealed that a lot of minor things have to pay attention during the use of AI in places related to public governance. Dharssini et al. [29] described the importance of smart meters, which are an example of IoT, AI, and DL. In the in-stitutional buildings, smart energy meters provide help in the estimation of exact energy requirements and to maintain a promising energy profile. Chen et al. [30] highlighted the benefits of DL in the manufacturing process of smart devices. It has been noticed that DL has major appli-cations in the manufacturing process of smart and intelligent devices. Alshdaifat et al. [2] highlighted the importance of machine learning algorithms to generate performance prediction models for students. The results of the study indicated that machine learning algorithms play a significant role in the development of prediction models.

16.3 ADVANTAGES OF AI, DL, AND IOT

Advantages		
AI	DL	IoT
Less possibility of human errors	Easily solve unstructured data	Time saving
High risk taking ability	Produce high-quality results	Security
24*7 workability	Cost-effectiveness	Less human efforts
No aggression to perform repetitive jobs	No need for data labelling	High data collection
Quick decision ability	Higher self-learning ability	Proper utilization of available resources
Unbiased	Support algorithms	High speed
Faster and smarter	Scalability	Easy access
High invention ability	Advanced analytics	Economic

16.4 LIMITATIONS OF AI, DL, AND IOT

Limitations		
AI	DL	IoT
High cost	Demand large amount of data	Data contravention
Emotionless	Requirement of deep knowledge	Complex system
Increasing unemployment	Results interpretation	Dependency upon Internet
Can't perform beyond the limits	Lack of flexibility	Absence of international standardization
Lack of creativity	Overfitting the model	Privacy issues

16.5 APPLICATIONS OF AI, DL, AND IOT IN DAILY LIFE

Applications		
AI	DL	IoT
Social media	Sentiment analysis	Agriculture
Navigation	Healthcare services	Smart cars
Travelling	Social media activities	Disaster management
Music	Virtual assistance	Home automation
Smartphones, etc.	Self-driving cars, etc.	Biometrical security system, etc.

16.6 FUTURE PERSPECTIVES OF AI, DL, AND IOT

It is very tough to estimate that in the future time period, when the recession may come in the technical field as the future is uncertain and nothing is permanent in the world [31]. AI is changing the scenario of every industry and impacting the life of every human being in the present era [c]. The uses of AI are increasing in robotics, healthcare services, smartphones applications, etc. DL has become a star overnight as a robot player defeated a human player during the famed game of Alpha Go. It is a trending AI-based technology which has offered benefits in various areas such as speech recognition, object recognition, language processing, etc. [32]. At present, a lot of companies are enjoying different kinds of operational and financial advantages due to DL [33]. So learning and training methods of DL are highly acknowledged for its importance and advantages for human beings [d]. The benefits of IoT can be highly noticed in various areas such as environmental monitoring, logistics, animal farming, healthcare, industrial control, etc. [e]. So on the basis of all advantages of AI, DL, and IoT, it can be concluded that the future of all these technical advancements is definitely bright. But it will produce positive results only in that case when it will be used for a positive purpose by ignoring its negative uses.

16.7 CONCLUSION

After the comparison of a lot of advantages, limitations, applications, and future perspectives of AI, DL, and IoT it has been found that all these types of the latest technology are boons for the world as a whole. Different kinds of advancements in various fields are the result of these latest technologies. But somewhere it has some limitations also. People should avoid the negative uses of these kinds of technology and try to focus on new advancements in such areas. Because all things depend upon human mentality, whether they pay more concentration to exploit the opportunities of technology or use technology in negative manners. The present study describes various aspects of AI, DL, and IoT in the form of their advantages and disadvantages. It will offer reference to the various researchers to move forward with their studies in their respective fields. After the analysis of the limitations of such described concepts, various measures can be generated to solve the issues.

REFERENCES

[1] Haldorai, A., Sharif, M., & Ravana, S. R. (2020). Artificial intelligence for sustainable internet research. *International Journal of Computers Applications*, available at www.researcher-app.com

[2] Alshdaifat, E., Zaid, A., & Alshdaifat, A. (2022). Predicting student performance: a classification model using machine learning algorithms. *International Journal of Business Information Systems*, available at www. researchgate.net

[3] Yashodha, G., Rani, P. Lavanya, A., & Sathyavathy, V. (2021). Role of artificial intelligence in the internet of things – a review. *IOP Conference Series: Materials Science and Engineering*, available at https://iopscience. iop.org

[4] Dimililer, K., Dindar, H., & Al-Turjman, F. (2021). Deep learning, machine learning and internet of things in geophysical engineering applications: an overview. *Microprocessors and Microsystems, 80*, available at www. semanticscholar.org

[5] Qinxia, H., Nazir, S., Li, M., Ullah Khan, H., Lianlian, W., & Sultan Ahmad, W. (2021). AI enabled sensing and decision-making for IoT systems. *Complexity*, available at www.hindawi.com

[6] Bala, R. (2021). Students' perspective of cloud-based digitalization of higher education. *Digitalization of Higher Education using Cloud Computing*, 1st ed., available at www.taylorfrancis.com

[7] Madakam, S., Ramaswamy, R., & Tripathi, S. (2015). Internet of things (IoT): a literature review. *Journal of Computer and Communications, 03*, available at www.scirp.org

[8] Osuwa, A., Ekhoragbon, E., & Fat, L. (2017). Application of artificial intelligence in internet of things. In *Proceedings of 9th International Conference on Computational Intelligence and Communication Networks (CICN)*, available at https://ieeexplore.ieee.org

[9] Sarmah, A., Baruah, K., & Baruah, A. (2017). A brief review on internet of things. *International Research Journal of Engineering and Technology (IRJET), 4*(10), available at https://www.irjet.net

[10] Ahmed, H. E. (2018). AI advantages & disadvantage. *International Journal of Scientific Engineering and Applied Science (IJSEAS)*, pp. 4(4), available at www.ijseas.com

[11] Boutaba, R., Salahuddin, M. A., Limam, N., Ayoubi, S., Shahriar, N., Estrada-Solano, F., & Caicedo, O. M. (2018). A comprehensive survey on machine learning for networking: evolution, applications and research opportunities. *Journal of Internet Service and Applications, 9*, available at https://jisajournal.springeropen.com

[12] Ghosh, A., Chakraborty, D., & Law, A. (2018). Artificial intelligence in internet of things. *CAAI Transactions on Intelligence Technology, 3*, available at www.researchgate.net

[13] Olley, D. (2018). Project report on artificial intelligence: how knowledge is created, transferred, and used trends in China, Europe, and the United States. pp. 1–92, available at https://www.elsevier.com

[14] Hussein, A. (2019). Internet of Things (IOT): research challenges and future applications. *International Journal of Advanced Computer Science and Applications, 10*, available at www.ijacsa.thesai.org

[15] Kankanhalli, A., Charalabidis, Y., & Mellouli, S. (2019). IoT and AI for smart government: a research agenda. *Government Information Quarterly, 36*, 304–309, available at https://www.sciencedirect.com

[16] Kumar, S., Tiwari, P., & Zymbler, M. (2019). Internet of things is a revolutionary approach for future technology enhancement: a review. *Journal of Big Data*, 6, available at https://journalofbigdata.springeropen.com

[17] Pal, H. (2019). Internet of Things (IOT): a study analysis of applications and benefits in health care sector. *International Journal of Engineering Research & Technology*, 8, available at www.ijert.org

[18] Sharma, K., & Nandal, R. (2019). A literature study on machine learning fusion with IoT. In *Proceedings of 3rd International Conference on Trends in Electronics and Informatics (ICOEI)*, available at https://ieeexplore. ieee.org

[19] Brous, P., Janssen, M., & Herder, P. (2020). The dual effects of the Internet of Things (IoT): a systematic review of the benefits and risks of IoT adoption by organizations. *International Journal of Information Management*, 51, available at www.sciencedirect.com

[20] Goyal, A., Nandi, A., Sharma, I., Sen, I., & Pathak, N. K. R. (2020). Project report on implementation of machine learning in IoT, available at https://jecassam.ac.in

[21] Merenda, M., Porcaro, C., & Iero, D. (2020). Edge machine learning for AI-enabled IoT devices: a review. *Sensors*, 20, available at www.mdpi.com.

[22] Mufti, T., Sami, N., Sohail, S. S., Siddiqui, J., Kumar, D., & Neha (2020). Future Internet of Things (IOT) from cloud perspective: aspects, applications and challenges. *Internet of Things (IoT)*, pp. 515–532, available at https://www.researchgate.net

[23] Bala, R. (2021). Privacy and ethical issues in digitalization world. *Internet of Things*, 1st ed., available at www.taylorfrancis.com

[24] Bolhasani, H., Mohseni, M., & Rahmani, A. (2021). Deep learning applications for IoT in health care: a systematic review. *Informatics in Medicine Unlocked*, 23, available at https://www.researchgate.net

[25] Kuzlu, M., Fair, C., & Guler, O. (2021). Role of artificial intelligence in the Internet of Things (IoT) cybersecurity. *Discover Internet Things*, 7, available at https://link.springer.com

[26] Sultana, N., & Tamanna, M. (2021). Exploring the benefits and challenges of Internet of Things (IoT) during Covid-19: a case study of bangladesh. *Discovery of Internet Things*, 20, available at https://link. springer.com

[27] Toorajipour, R., Sohrabpour, V., Nazarpour, A., Oghazi, P., & Fischl, M. (2021). Artificial intelligence in supply chain management: a systematic literature review. *Journal of Business Research*, 122, 502–517, available at www.sciencedirect.com

[28] Zuiderwijk, A., Chen, Y., & Salem, F. (2021). Implications of the use of artificial intelligence in public governance: a systematic literature review and a research agenda. *Government Information Quarterly*, 38, available at www.sciencedirect.com

[29] Dharssini, V., Raja, C., & Karthick, T. (2022). Benefits of smart meters in institutional building – a case study. *Smart Buildings Digitalization*, 1st ed., available at https://www.taylorfrancis.com

[30] Chen, M., Lughofer, E. D., & Egrioglu, E. (2022). Deep learning and intelligent system towards smart manufacturing. *Enterprise Information Systems*, 16, 189–192, available at www.tandfonline.com

[31] Torresen, J. (2014). Future perspectives on artificial intelligence (AI), available at https://www.uio.no

[32] Rajan, V., & Hepziba, G. (2017). Perspectives and future outlook of deep learning AI. *International Journal of Modern Trends in Engineering and Research (IJMTER)*, 4, 86–94, available at https://www.researchgate.net

[33] Bellapu, A. (2021). The future of deep learning, available at https://www.analyticsinsight.net

Chapter 17

A Support Vector Machine–Based Deep Learning Approach for Object Detection

Vishal Gupta and Monish Gupta

CONTENTS

17.1 INTRODUCTION

The small objects are very hard to detect in the worst weather conditions which might interfere with the low-intensity pixel values. The area, a fundamental development in video assessment, is dire in application districts such as motorized visual observation, human–PC joint effort, the frames extracted from the video processing, and modified traffic noticing [1–4]. All things considered, each following computation requires a thing ID instrument to perceive fights either in each packaging while frame extraction from the video processing [5,6], and the accompanying show of visual surveillance structures is dependent upon the sufficiency of article revelation. Moving thing recognizable proof isolates the interest in moving objects, for instance, vehicles and walkers in a video game plan with a static or dynamic establishment. Object disclosure is known as a task that tracks down all spots of objects of interest in a commitment by bobbing boxes and naming them into arrangements that they have a spot with. To do this task, a couple of considerations have been proposed from customary ways of managing significant learning-based approaches. The methodologies of thing area are generally segregated into two sorts, specifically, approaches considering region recommendation calculations are done using two-stage approaches [7–10] and approaches considering backslide or classification seen as one-stage approaches or non-stop and unified associations [8–11]. People are currently thinking about applications that take into account consistent

DOI: 10.1201/9781003355960-17

thing disclosure because of their income for meeting the state-of-the-art life and assisting people in having an unequalled life. Self-driving vehicles, for example, are a certified way to assist people to get around securely on the streets while reducing car accidents caused by distracted drivers [12]. The other recalls the necessity for collecting organizations to perceive missing or weak get-together elements, as well as the size of recognized objects and malleable shape that significantly varies during the social affair process [13–15]. It depicts that continuous object recognition, as applied to the most well-known video frame extraction-based applications in the certified world, is.

17.2 LITERATURE

The parted-based method is better for short-term tracking and has more flexible updating mechanisms than the holistic approach. These methods are usually reliable in tracking objects with motion fluctuations, and they can effectively handle tracking duties for objects with partial occlusion. The parted-based approach, on the other hand, is prone to drifting in the presence of backdrop clutter and motion blur because it ignores the entire visual information of the target. Nevertheless, such applications require early article areas to be used in this manner as commitments for various endeavours [15–20]. As a result of the early period, portrayal of items is for the most part minimal or even negligible. The justification for little article ID, given an image of interest, is to instantly recognize what ordinary things have a spot that is true with a small appearance (mouse, plate, compartment, bottle, etc.) [21]. As a result, a small article area is a difficult assignment in PC vision because it is segregated from small portrayals of things; the variety of data pictures also makes the task more difficult [22]. Dynamically, traffic observation systems, moving thing distinguishing proof taking into account pictures got from fixed CCTV cameras and other possible sources.

In dynamically introduced structures with insufficient computing and memory resources, the computational complexity of various estimations, such as object recognition, is also a difficult task [23]. As a result, object revelation estimations for embedded systems should be quick, straightforward, and reasonable, with commanding execution. We provide a thing revelation computation suitable for introduced perception applications in this letter. The proposed technique combines the genomic exceptional saliency map (GDSM) with the establishment allowance. The dynamic saliency map (DSM) is used in GDSM, which requires less estimation, has higher article area accuracy, and is more resistant to noise and environmental variations. The visual enhancement of GDSM over the standard DSM is due to the use of genetic computation to advance the heaps, whereas uniform burdens are employed in the standard DSM for creating a saliency map. In any case, GDSM fails to notice dissents, particularly when they abruptly halt in the middle of the road. As a result, we combine GDSM with establishment allowance (BS) to distinguish moving objects even more precisely. More stringent article restrictions and including areas are recognized by BS. We use running Gaussian averaging, which is known to be fast and direct, as one of the possible BS techniques [24–26].

A wide assortment of identification strategies has been proposed somewhat recently for the improvement of profound learning. Different thoughts have been introduced, and joined assessments have been made to manage difficulties of article recognition, however, those proposed identifiers as of now spend their capacity on the discovery of ordinary sizes, not simply little items. Nonetheless, an assessment of little item location approaches is vital and significant in the investigation of article recognition. Recently, object location has significantly stood out from best-in-class draws near, and these have made their efforts to handle object discovery and yield great execution on testing and multiclass are examples of datasets. These cutting-edge techniques are first prepared on ImageNet before being deployed; for example, instead of using part-based models, the creators use a proposed network that applies a spatial pyramid pooling layer to remove highlights and compute these over a whole picture without regard to picture sizes [27]. R-CNN [1] is a forerunner in cutting-edge object recognition, with a few improvements over previous methods: an image matrix pixel is resized to a fixed size to fit into the methodical parameter, and then an outer calculation is applied to produce item suggestions. Support vector machine–convolutional neural network (SVM–CNN) [28] improves by using regions of interest (ROI) to remove a fixed-length inclusion from each element map. Until now, nearly recognition models have primarily been used to evaluate datasets such as CIFAR. These datasets typically comprise things that take up medium or large parts of an image that contains a few little articles, resulting in an unevenness of information across objects of different sizes, causing models to gravitate toward objects with higher numbers. Furthermore, the current little item datasets have a different number of classes than typical datasets. Furthermore, the bulk of the finest in class locators has struggled to recognize little objects, both in one-stage and two-stage draws close. As a result, in our previous work [29], we introduced a top-to-bottom examination of existing profound learning models in discriminating small things. In this way and age, we not only expand by constantly evaluating best-in-class and new discovery models, but we also summarize advantages and disadvantages, as well as model plans, rather than simply presenting their ideas. Rather than focusing on fixed models, we evaluate cutting-edge models in two ways: one-stage approaches that can run continuously, such as YOLOv3 and Retina Net, and two-stage approaches that don't satisfy ongoing location but have excellent. Despite the fact that Faster R-CNN is the only model that has been evaluated in our previous work, we still need to test it with different spines to see how well they operate when combined with SVM–R-CNN. This model uses an outside calculation to build locality proposition on an information picture rather than on an element map. Furthermore, we evaluate these models with several open available datasets [30,31]. We evaluate two datasets, in particular, the small item openly available dataset used for the experimental purpose. Using standards such as precision, speed of processing, and asset usage. Regardless, we must examine the plan and the manner in which models operate.

17.3 PROPOSED METHODS

The proposed procedure is roused by the profound learning approach. There are many exploration works previously finished for the item recognitions utilizing various layers of convolutional layers and in the current work, we have proposed the straightforward profound learning 3x3 convolutional layers followed by the pooling layer. The developed methodology Quicker R-CNN [15] is an improvement on SVM–CNN. The previous techniques used for the bounding box may include the densed cluttered involved with the other object. This may lead to false detection and overfitting. RPN is considered after obtaining deep highlights from early convolutional layers, and windows glide over the element guide to disentangle highlights for each locale proposition. RPN is a totally convolutional network that predicts bouncing boxes of items as well as objectness scores at each place. (RPN contribution is an image of any size that yields a collection of rectangular item propositions as well as an objectness score for each proposition.) The RPN, in particular, uses different levels of processing (see Figure 17.1). The novelty of the work comes from the extracted features are fed to the convolutional layer for deep learning and further processing. By sharing their convolutional highlights, RPN and SVM–CNN are combined into a single organization. (This mix allows a faster R-CNN to have precise driving execution, but also forces it to be engineered as a two-stage network, slowing down the technique's handling speed.) The SVM is first able to distinguish between the background part from the desired part and extract the pixel based on the intensity level. Usually, the high-intensity pixels represent the edges of the objects and the common low-intensity pixels are considered as the noise. Since small objects may be difficult to distinguish from the noise factors as they might have pixels with low-intensity values. The extracted features are collected as a bag of bunch and are useful for the training of the convolutional layers.

The dictionary should be impervious to background distractions. Based on similarity, picture patches corresponding to differences in an object's appearance can be grouped into multiple clusters. Changes in viewpoint

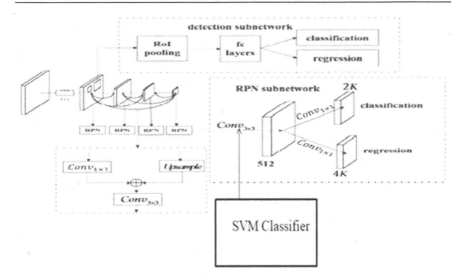

Figure 17.1 Proposed methodology.

and shape deformations, for example, both produce image patches that correspond to various views of a target or its background. As a result, picture patch distribution is rarely unimodal. A multimodal distribution, on the other hand, is more generalizable in terms of its structure. The complexity of such a structure, on the other hand, might easily lead to confusion in the selection of training samples, lowering the dictionary's quality. Sparse representation is also hampered by this difficulty.

In essence, SSD is divided into two sections: component guidance extraction and object identification using convolutional channels. VGG16 is used as a base structure to segregate elements in SSD. (It forecasts by combining six convolutional layers.) Every expectation has a bouncing box and obtained scores for each class, where N is the number of classes and one is for extraclass, which means there is no article. SSD uses small convolution channels instead of a district proposition structure to construct boxes and feed them to a classifier for registering the item area and class scores. SSD utilizes 3×3 convolution channels for each cell to predict items after the VGG16-based organization extracts highlights from inclusion maps. It is necessary to examine the short films in spatiotemporal space in order to construct a more efficient region proposal method with fewer candidate regions. As a result, we created a limited collection of region recommendations for the camera trap images that were processed using an Iterative Embedded-Graph Cut (IEC) approach. Detection of objects. Similarly, CNN was employed to improve region classification. A feature extractor was used, and the classification was done with the best-in-class classifiers SVM. In this chapter, we used two well-known CNN models, namely, the optical flow approach is also used for object detection in image sequences. Optical flow calculation of a picture requires clustering for the distribution of the optical flow of the image in the optical flow technique. The optical flow technique is used to extract the comprehensive information about an object based on its mobility and makes use of it to recognize the object. The reduction of false alarms and the exact detection of objects targets. During the beginning, we employ a Random-Forest-based clustering technique which assesses that the ROI are extracted in accordance with the saliency map and is used to remove false alarms in detail, whereas the other areas are left unprocessed, which considerably reduces the workload and the cost of calculation. In the second stage, we create a dynamic environment. A contour model (ACM) with dynamic background based on CFAR from each extracted ROI was used to predict the ship outline. Starting with the immediate environment. The CNN network geographically reduces the component of the image over time, resulting in a reduction in the element maps' aim. SSD, as previously said, recognizes objects using a lower input image; as a result, early layers are used to detect small articles, while lower goal layers are used to logically discern larger scope protests. SSD also applies various sizes of default boxes to various levels, for intuitive depiction. Specifically, the major blue default box on the 8 × 8 component map corresponds to the feline's ground reality, while the main red default box on the 4 × 4 element map corresponds to the canine's ground reality. On PASCAL VOC and COCO, SSD outperforms the proposed model, which is a best-in-class technique in terms of exactness, for the most part, while operating at ongoing recognition.

17.4 RESULTS

This section concludes the results that we accomplished through the trial stage. All models referenced in this segment with the exception of models referred to from different papers are prepared on a similar climate MATLAB 2021 R. Notwithstanding the similar precision, different correlations are likewise given to make our goal clear appraisal results.

Informational index 1 is gathered under shady climate, informational index 2 under blustery climate, and informational collection 3 in bright days, with shadows. On account of informational collection 3, because of the stains on the camera by the downpour, some identification execution corruption is noticed. The model shows the aftereffect of the shadow concealment, showing that the shadows can be successfully recognized and eliminated by the proposed technique as shown in Figures 17.2 and Figures 17.3, respectively. Albeit a few pixels are erroneously identified as shadows, the proposed technique is effective in identifying the article district, as well as impediments in each casing. When the impediment is distinguished, the blocked items are looked at

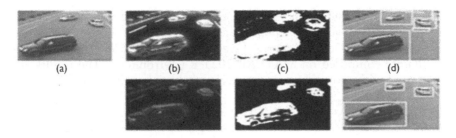

Figure 17.2 Input Image processing and output.

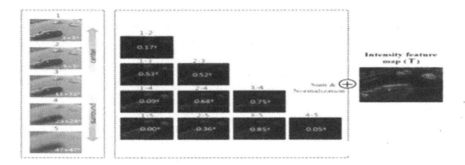

Figure 17.3 Output.

Table 17.1 Methods

Parameters	Accuracy (%)	False ratio (%)	Computing time (sec)
Proposed method	87.87	5.6	3.6
Previous method [10]	84.56	9.85	8.9

with the mean-shift calculation as displayed. In this way, the proposed strategy can distinguish and recognize numerous items, not at all like the regular techniques that recognize every one of the impeded articles as a solitary item.

Table 17.1 shows the calculated results which show the accuracy in detecting the small object which is obtained while reducing the computing time. The proposed model is capable to detect the desired targets by neglecting the noise facts, hence reducing the false ratio.

17.5 CONCLUSION

The proposed methodology is well defined and presented. The deep features are extracted using the SVM and then the model is trained.

The small object detection technique is able to extract the desired features which results in achieving higher accuracy. The extracted features are used to train the convolutional layer of the model. The mathematical results are computed on MATLAB 2021 R environment. The results are compared with the previous approach and show better accuracy.

REFERENCES

[1] Kyung Hee, M. (May 2007). A dynamic histogram equalization for image contrast enhancement. *IEEE Transaction on Consumer Electronics*, *53*(2), 593–600.

[2] Kim (Feb. 1997). Contrast enhancement using brightness preserving bi-histogram equalization. *IEEE Transaction on Consumer Electronics*, *43*(1), 1–8.

[3] Wan, Y., Chen, Q., & Zhang, B. (Feb. 1999). Image enhancement based on equal area dualistic sub-image histogram equalization method. *IEEE Transaction Consumer Electronics*, *45*(1), 68–75.

[4] Ramli (2009). Minimum mean brightness error bi-histogram equalization in contrast enhancement. *IEEE Transaction Consumer Electronics*, *49*(4), 1310–1319.

[5] Zhang, J., Ehinger, K. A., Wei, H., Zhang, K., & Yang, J. (2017). A novel graph-based optimization framework for salient object detection. *Pattern Recognition*, *64*, 39–50.

[6] Wang, & Ye, Z. (Nov. 2005). Brightness preserving histogram equalization with maximum entropy: a variation perspective. *IEEE Transaction on Consumer Electronics*, *51*(4), 1326–1334.

[7] Ibrahim (2007). Brightness preserving dynamic histogram equalization for image contrast enhancement. *IEEE Transaction on Consumer Electronics*, *53*(4), 1752–1758.

[8] Lamberti, F., Montrucchio, B., & Sanna, A. (Aug. 2006). CMBFHE: a novel contrast enhancement technique based on cascaded multistep binomial filtering histogram equalization. *IEEE Transaction on Consumer Electronics*, *52*(3), 966–974.

[9] Kim, L.-S. (2001). Partially overlapped sub-block histogram equalization. *IEEE Transaction on Circuits and Systems for Video Technology*, *11*(4), 475–484.

[10] Celik, T., & Tjahadi, T. (Dec. 2011). Contextual and variation contrast enhancement. *IEEE Transaction on Image Processing Applications*, *20*(2), 3431–3441.

[11] Fu, K., Gu, I. Y.-H., & Yang, J. (2018). Spectral salient object detection. *Neurocomputing*, *275*, 788–803.

[12] Rodriguez Sullivan, M., & Shah, M. (2008). Visual surveillance in maritime port facilities. *Proceedings of SPIE*, *6978*, 29.

[13] Liu, H., Javed, O., Taylor, G., Cao, X., & Haering, N. (2008). Omni-directional surveillance for unmanned water vehicles. In *Proceedings of International Workshop on Visual Surveillance*. USA.

[14] Wei, H., Nguyen, H., Ramu, P., Raju, C., Liu, X., & Yadegar, J. (2009). Automated intelligent video surveillance system for ships. *Proceedings of SPIE*, 7306, 73061N.

[15] Fefilatyev, S., Goldgof, D., & Lembke, C. (2009). Autonomous buoy platform for low-cost visual maritime surveillance: design and initial deployment. *Proceedings of SPIE*, 7317, 73170A.

[16] Kruger, W., & Orlov, Z. (2010). Robust layer-based boat detection and multi-target-tracking in maritime environments. In *Proceedings of International Waterside*. USA.

[17] Fefilatyev, S., Shreve, M., & Lembke, C. (2012). Detection and tracking of ships in open sea with rapidly moving buoy-mounted camera system. *Ocean Engineering*, 54, 1–12.

[18] Westall, P., Ford, J., O'Shea, P., & Hrabar, S. (Dec. 2008). Evaluation of machine vision techniques for aerial search of humans in maritime environments. In *Digital Image Computing: Techniques and Applications (DICTA) 2008 (Canberra, 1–3)*, Vol. 29, pp. 176–183.

[19] Herselman, P. L., Baker, C. J., & Wind, de H. J. (2008). An analysis of X-band calibrated sea clutter and small boat reflectivity at medium-to-low grazing angles. *International Journal of Navigation Observation*, 10.1155/2 008/347518

[20] Gupta, V., & Gupta, M. (2020). Ships classification using neural network based on radar scattering. *International Journal of Advanced Science and Technology*, 29, 1349–1354.

[21] Fu, K., Gu, I. Y.-H., & Yang, J. (2018). Spectral salient object detection. *Neurocomputing*, 275, 788–803.

[22] Pathak, A. R., Pandey, M., & Rautaray, S. (2018). Application of deep learning for object detection. *Procedia Computer Science*, 32, 1706–1717.

[23] Gupta, V., & Gupta, M. Automated object detection system in marine environment. *Mobile Radio Communications and 5G Networks, Lecture Notes in Networks and Systems*, 140, 10.1007/978-981-15-7130-5_17

[24] Tu, Z., Guo, Z., Xie, W., Yan, M., & Yuan, J. (2017). Fusing disparate object signatures for salient object detection in video. *Pattern Recognition*, 72, 285–299.

[25] Goyal, K., & Singhai, J. (2018). Texture-based self-adaptive moving object detection technique for complex scenes. *Computers & Electrical Engineering*, 70, 275–283.

[26] Hou, S., Wang, Z., & Wu, F. (2018). Object detection via deeply exploiting depth information. *Neurocomputing*, 286, 58–66.

[27] Wang, G., Zhang, Y., & Li, J. (2017). High-level background prior based salient object detection. *Journal of Visual Communication and Image Representation*, 48, 432–441.

[28] Gupta, V., Gupta, M., & Singla, P. (2021). Ship detection from highly cluttered images using convolutional neural network. *Wireless Personal Communication*, 10.1007/s11277-021-08635-5. Springer.

[29] Gupta, V., Gupta, M., & Marriwala, N. A modified weighed histogram approach for image enhancement using optimized alpha parameter. *Mobile Radio Communications & 5G Networks (MRCN–2021), Lecture Notes in Networks and Systems*, 10.1007/978-981-16-7018-3_39, Springer Conference.

[30] Gupta, V., & Gupta, M. (2021). IoT based artificial intelligence system in object detection. *Internet of things: energy, industry and healthcare.* CRC Press, Taylor and Francis Group, USA.

[31] Gupta, V., Marriwala, N., & Gupta, M. (2021), A GUI based application for low intensity object classification & count using SVM approach. In *2021 6th International Conference on Signal Processing, Computing and Control (ISPCC)* (pp. 299–302). 10.1109/ISPCC53510.2021.9609470. IEEE.

Chapter 18

Artificial Intelligence and IOT-Based Electric Grid: Future of Micro Grid

Dr. Cheshta Jain Khare, Dr. Vikas Khare, and Dr. H.K. Verma

CONTENTS

18.1 INTRODUCTION

The power system exhibits traits of openness, ambiguity, and complexity as a result of electricity market reform and renewable energy and power demand response application scenarios. The development and implementation of smart power grids have become fashionable. Artificial intelligence (AI) in smart grid applications provides substantial technological provision for digital power networks [1–10].

Rising consumption, efficiency, shifting supply and demand patterns, and a lack of analytics necessary for optimal management are all difficulties facing the energy sector around the world. Emerging market countries are particularly vulnerable to these concerns. AI and related technologies for communication between smart grids, smart meters, and Internet of Thing (IOT) devices are already being used in the power sector in developed countries. AI is rapidly being used in the power sector, and it has the potential to have a large impact in emerging economies, where clean, inexpensive, and dependable electricity is critical for growth.

DOI: 10.1201/9781003355960-18

AI has the ability to minimize wastage of energy, inferior energy costs, and increase the adoption of clean non-conventional energy sources in global power networks. AI can help with power system planning, operation, and control. As a result, AI is intrinsically related to the capability to offer clean, affordable energy, which is essential for development.

The electrical business has a bright future ahead of it, thanks to solutions such as AI-managed smart grids. These are two-way electrical networks that allow utilities and consumers to communicate. Smart grids feature an information layer that permits communication between their many components, allowing them to respond faster to unexpected changes in energy demand or emergencies. This data layer, which was formed by the extensive installation of smart meters and sensors, allows for data gathering, storage, and analysis. The IOTs smart grid, on the other hand, enables two-way communication between connected devices and hardware that detects and responds to human needs. Smart grids are more dependable and cost less than current electrical infrastructure.

These smart grid components have facilitated to increase the reliability, safety, and efficiency of energy transmission and distribution networks when combined with strong data analytics. AI technologies such as machine learning are appropriate for analyzing and interpreting such data due to the huge amount and variety of designs. These data analyses can be used for a variety of purposes, such as detecting problems, doing predictive maintenance, monitoring power quality, and forecasting renewable energy. Because smart grids are designed to reduce peak loads on distribution feeds. For more sustainable solutions, the US Department of Energy is incorporating green technology into their IOT smart management. All distribution networks could benefit from these solutions, which include the following:

- Wind turbines that are more efficient
- Photovoltaic cells
- Technology based on microgrids
- Automated feeder systems

The proliferation of smart meters has been aided by advances in information and communication technology (ICT), cloud computing, big data analytics, and AI.

18.2 MODELLING OF AI-BASED ELECTRIC GRID

The smart grid is a concept that entails switching from an electro-mechanically managed to an electronically controlled electric power infra-structure. In at least three significant ways, smart grid technologies have impacted traditional grid planning and operating challenges. The most important of which is the capacity to predict and avoid outages.

1. Track or extent processes, send data back to operations centres, and do so often respond repeatedly to adjust a process.
2. Communicate data between devices and systems; and
3. Process, analyze, and assist operators in accessing and applying digital data technology all throughout the grid.

18.2.1 Features of Smart Grid

Smart grid is defined variously by different countries and organizations, although the connotations of the various definitions are essentially the same. The smart grid is a system of highly automated power transmission nodes that allows data and energy to flow back and forth. Smart grids have better performance and can offer a variety of value-added services to users. The following are characteristics of an ideal smart grid.

- The electricity grid can self-repair. Self-repair control systems in smart grids may automatically identify and improve, swiftly restore power supply, reduce power outage time, and decrease power outage frequency and range.
- The operation is efficient. Smart grid can help power grid companies maximise their asset allocation. Dynamic review and modification, for example, can help organizations get the most out of their assets and enhance asset utilization efficiency.
- The electrical market should be improved. The growth of the electrical sector is dependent on good infrastructure and technological support. The smart grid ensures the operation of the electrical market by connecting numerous buyers and sellers with innovative technology and information systems.

The electric grid consists of a network of power plants and transmission lines that transport electricity to clients.

With the fast development of the population and the number of industries, managing the demand for electricity for home and industrial reasons has become a demanding problem for the networks. The increased demand for power at certain times of the day resulted in a number of issues, including short circuits and transformer breakdown. To overcome these challenges of electricity transmission through traditional grids, it is necessary to forecast consumer usage patterns in order to successfully supply electricity. The notion of smart grids (SG) has been introduced in this context. A smart grid can intelligently estimate electricity consumption and, as a result, transmit electricity depending on that need. An SG, by its intelligent sensing and prediction, may address various issues that plague traditional grids, including demand forecasting, power consumption reduction, and short circuit danger reduction, all of which save lives and property.

According to digital technology, "grid" refers to the electric grid, which is a bi-directional network communication line connecting consumers and utilities. Electricity outages are becoming more regular, resulting in a slew of failures in industries such as banking, communications, traffic, and security. Smart grid technology is used for this goal, and it is based on a variety of AI approaches. This scheme allows for the supervision, investigation, organization, and sharing of information throughout the supply chain in order to enhance efficiency, decrease energy utilization and costs, and make the most of the energy supply chain's transparency and reliability. The smart grid system addresses the shortcomings of traditional electrical grids, where the smart net meter concept was previously deployed. The concept of a smart grid strategy focuses not only on utilities and technologies but also on customer choices around electricity energy usage. Figure 18.1 shows the layout of a smart grid.

18.2.2 AI and Its Application to Smart Grid

Traditional modelling, optimization, and control methods, on the other hand, have significant boundaries in handling these datasets, necessitating the use of AI approaches in the smart grid.

AI can be expressed as computers that mimic grid operators' cognitive processes in order to achieve self-healing capabilities as smart grid applications. In other circumstances, though, AI might not be capable to fully

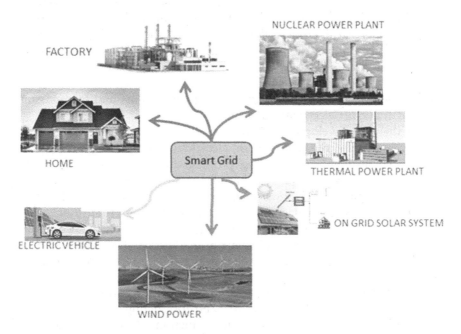

Figure 18.1 Layout of smart grid.

change grid operators. Despite the fact that AI systems can be more exact, dependable, and comprehensive, there are still a number of obstacles to overcome when incorporating AI into the smart grid. In the smart grid, virtual AI and physical AI are the two types of AI systems. Informatics in virtual AI systems can be beneficial to grid operators. Physical AI systems are self-aware AI systems that, with or without human intervention, can optimize and control particular grid processes. The AI approaches used in the smart grid are divided into the following categories.

ES: An expert in the loop technique, which is utilized to solve a variety of issues.

Supervised learning is a type of AI in which inputs and outputs are correlated and explored in order to anticipate the outputs of fresh inputs.

Unsupervised learning is a type of machine learning in which unlabelled data are utilized to capture data similarities and differences.

Reinforcement learning (RL) is distinguished from supervised and unsupervised learning by its intelligent agent's method, which tries to maximize the concept of cumulative learning.

Ensemble approaches combine the results of numerous AI algorithms to overcome the limits of a single algorithm while improving overall performance.

18.2.2.1 Application in Smart Grid

a. Load forecasting: Forecasting power load allows for real-time matching of power output and load, which is a daily task for power grid operators. The technique of estimating power demand in the system is known as electricity load forecasting. The three categories of power load forecasting are short-, medium-, and long-term, with time intervals ranging from a few minutes to over a year. Artificial neural network technology has been one of the most widely utilized ways for predicting power load in recent years. In its early phases, the artificial neural network's backpropagation method (BP algorithm) was a model with only one hidden node. The model can't accurately explain complex functions when there are only a few samples and computational elements. As technology progresses, deep learning is increasingly being used in power system load predictions. The long and short time memory (LSTM) network is a well-known deep learning paradigm. This is a more advanced recursive neural network–based model (RNN). Memory modules, rather than common hidden nodes, are used in LSTM networks to ensure that the gradient does not disappear or extend after several time steps, eliminating some of the issues that can occur with traditional RNN training.

b. Consumer electricity consumption: Machine learning clustering and identification AI talents could be used to analyse consumer power consumption patterns, detect anomalous power usage, and perform non-invasive load monitoring. These analyses and testing provide a theoretical basis for reasonable comprehensive energy system pricing and energy structure enhancement, as well as two-way flexible interaction between energy supply and users.

c. Network security: Real-time perception, information services, and dynamic control are all part of the smart grid. As a result of the deep information flow interaction, the power system will be exposed to more hazards. The network assault on the power system is well-hidden and takes a long time to develop. The secondary system can be damaged to attack the physical power grid, even if the primary equipment is not immediately harmed. Deep learning can detect malware and intrusion, as well as provide network security protection for power systems, by automatically recognizing network attack features. Anomaly sample data from an attacked power network is much lower than normal sample data because the chances of a power system being attacked are much lower than in normal operation.

d. Renewable energy power generation is gaining popularity, however, the intermittent and erratic nature of renewable energy generation will have an influence on system stability. The power system's steady, efficient, and cost-effective operation depends on the precise prediction of renewable energy generation. The classic shallow model prediction method performs poorly when dealing with nonlinear and non-stationary wind or light data. The LSTM model, which is based on the load prediction concept, may also be used to estimate the electricity generated by wind and photovoltaic panels.

e. Losses caused by unauthorized connections are another source of concern for the power industry. AI might be used to spot discrepancies in consumption patterns, payment histories, and other consumer data in order to detect these unofficial links. When used in conjunction with automated meters, it can also improve monitoring for them.

18.2.2.2 Overview and Implementation of Deep Learning Algorithm

By merging modern technologies such as ML, IOT, and big data analytics with SG, clients' power demand may be forecasted and demand response can be automated. Deep learning (DL) based models can be used to find patterns in the data and anticipate electricity demand during peak hours because the SG network creates a significant volume of data.

Due to the massive amount of data received from various sources at all times, it is vital for all power companies to accept responsibility for all obtained data and gain a comprehensive understanding of electricity consumption behaviours, as indicated in the block diagram (Figure 18.2). For

Renewable Energy

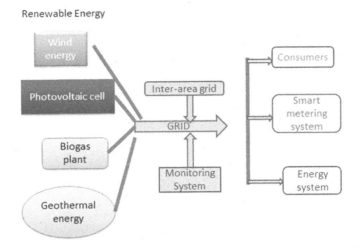

Figure 18.2 Block diagram of electricity consumption.

feature extraction and data processing, a deep learning algorithm is necessary. The majority of the investigations, such as load outlines, customer classification, load forecasting, and anomaly detection, were conducted using smart meter data. Flexible demand management and meaningful energy control smart meter data analysis are required for a better understanding of electricity use.

Deep learning makes use of vast neural networks with several layers of processing units to learn a variety of patterns from a large amount of data with improved computational power and training methodologies. Image and speech recognition are two common uses. Neural networks are used to implement deep learning. The biological neuron provides the inspiration for neural networks. The basic structure for prediction of output using feed-forward ANN is shown in Figure 18.3

The network shown has one hidden layer, input layer, and one output. Where

Xin = number of inputs
Im = number of input layer neurons
hn = number of hidden layer neurons
w(Im,hn) = weights of input layer from Imth neuron to hnth neuron.
O(hn,1) = weights of hidden layer from hnth neuron to output.

The input layer neurons simply distribute signals, but the hidden layer neurons are activated by a sigmoidal (logistic) function. Finally, a linear function activates the neurons in the output layer. The steepest-descent minimization algorithm is a variation of the learning algorithm used for training. Two structures were created for the aim of prediction.

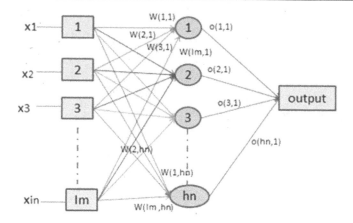

Figure 18.3 Basic structure of an artificial neural network.

Time-controlled recurrent (TCR) was the name of the first one. Figure 18.4 depicts it. Feed-forward accommodated for prediction (FFAP) was the second one, which is shown in Figure 18.5.

It's worth noting that, deterministic forecasting requires at least two forecasts that are mutually supported. Because no information about the predicted outcome is available, the second prediction is merely a means of confirming the first. However, because both predictions have the same level of uncertainty, we chose to accept the average of the two as the best final prediction. Predicting electric load is critical for power generation and

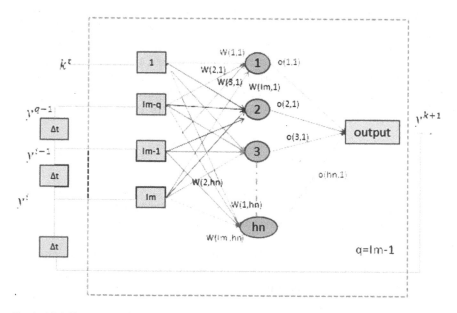

Figure 18.4 Structure of time-controlled recurrent (TCR).

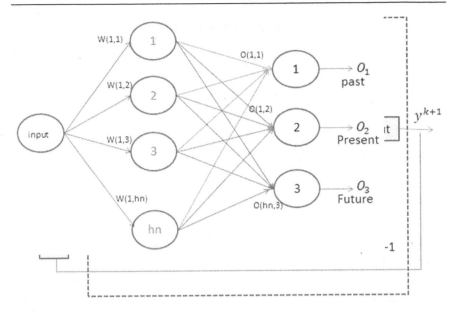

Figure 18.5 Structure of feed-forward accommodation for prediction algorithm.

operation. It's critical in a variety of ways, including cost-effective generating, system security, and planning. It allows for the scheduling of fuel purchases, power generation, energy transaction planning, and system safety assessment, among other things. The additional expenditures incurred as a result of load forecast inaccuracies are significant: if the load is understated, network investment costs exceed real needs, and fuel reserves are overvalued, capital investment is locked up. If the load is overstated, network investment costs exceed real needs, and fuel reserves are overvalued, capital investment is locked up. Prediction in the context of a smart grid allows for the development of computationally efficient learning algorithms that can accurately predict both the prosumers' (producer/consumer) consumption and generation profiles (rather than just the consumer's usage profile), as well as the price of electricity in real-time, allowing for profitable trading decisions.

18.2.2.2.1 Implementation

To implement a deep learning algorithm, the following steps may be followed:

Step 1. Obtain the peak load of the prior days and guess the peak load for the future using the daily load curve. Based on these daily peak loads, draw the monthly load curve. For example, load curve of May month is shown in Figure 18.6.

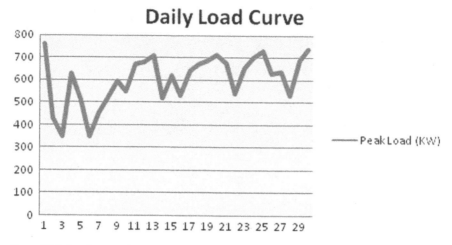

Figure 18.6 Load curve for the month of May.

Step 2. Now define the problem as follows: define series as $(d_k, f(d_k))$ for $k = 1, 2, \ldots nn$; where d_k is the day in the calendar, $f(d_k)$ the peak-to-peak value at that day. Make the series up to last peak value at nnth day. Our aim is to predict the peak value at $(nn + 1)$th day.

Step 3. Now develop ANN structure which may be time-controlled recurrent based or feed-forward accommodated for prediction based.

Step 4. If one wishes to have any faith in a prognosis, he or she must have at least two predictions that are mutually supportive. Because both are equally important, the average is calculated and stated as the final result instead of accepting one of them.

18.2.2.2.2 Solution Using Extended Feed-forward Accommodated for Prediction

In the feed-forward ANN, input layers have data from peak values of previous weeks and the current day. The output layer is the function of peak values of the shift in time for one day. For example, design structure for prediction of load for June month then inputs are the weekly peak loads of May month. We can take three or five inputs according to available data. Now, consider five outputs which are defined as:

$$\{y_{d+1}, y_d, y_{d-1}, y_{d-2}, y_{d-3}\} = f(d, y_{d-5}, y_{d-12}) \tag{18.1}$$

where y_d is the peak load at day d.

To implement the algorithm, the key element is the activation function. Each node in the preceding layer is connected to all the nodes in the layer before it. Each connection is assigned a weight, after which the node sums all of the incoming inputs multiplied by the connection weight. The total of these inputs is added together and used as an input to an activation function.

The activation function is a mathematical formula that determines whether a node has sufficient informative input to shoot a signal to the next layer. There are a variety of activation functions to choose from, but the following are the most common:

$$\text{linear: } f(d) = d \tag{18.2}$$

$$\text{Rectified linear unit: } f(d) = \begin{cases} 0 \text{ for } d < 0 \\ d \text{ for } d \geq 0 \end{cases} \tag{18.3}$$

$$\text{sigmoid: } f(d) = \frac{1}{1 + e^{-d}} \tag{18.4}$$

Training algorithm:

Step 1. Initialize weights of input and hidden layers and learning rate.
Step 2. Repeat steps 3–10 till stopping criterion is not satisfied.
Step 3. Repeat steps 4–9 for each neurons.
Step 4. Each input neuron sends a signal to the neurons of the hidden layer ($i = 1$ to m)
Step 5. Each hidden neuron ($j = 1$ to n) sums weighted inputs and obtains net input as:

$$h_{inj} = \sum_{i=1}^{m} I_i w_{ij} \tag{18.5}$$

Now calculate the output of the hidden neuron by using the appropriate activation function of h_{inj}:

$$h_j = f(h_{inj}) \tag{18.6}$$

Send this output from hidden neuron to the output neuron.

Step 6. For each output neuron determine the net input:

$$y_{ink} = \sum_{j=1}^{n} h_j o_{jk} \tag{18.7}$$

and apply activation function and get outputs:

$$y_k = f(y_{ink}) \tag{18.8}$$

Step 7. Each output ($k = 1$ to nn) has a desired output corresponding to input then compute the error correction as:

$$e_k = (\text{Desired output}_k - y_k)f'(y_{ink}) \tag{18.9}$$

The first derivative of function (f') is

$$f'(y_{ink}) = \lambda f(y_{ink})(1 - f(y_{ink}));$$

where λ is the steepness parameter.
  Using this error update weights as

$$\Delta o_{jk} = \alpha e_k h_j; \text{ where } \alpha \text{ is learning rate parameter} \tag{18.10}$$

Step 8. Each hidden neuron sums its updated inputs from outputs as:

$$e_{inj} = \sum_{k=1}^{nn} e_k o_{jk} \tag{18.11}$$

$$e_j = e_{inj} f'(h_{inj}) \tag{18.12}$$

Using e_j update weights of input layer:

$$\Delta w_{ij} = \alpha e_j I_m \tag{18.13}$$

Step 9. Each output neuron update the weights:

$$o_{ij}(\text{new}) = o_{ij}(\text{old}) + \Delta o_{ij} \tag{18.14}$$

Hidden neurons also update their weights:

$$w_{ij}(\text{new}) = w_{ij}(\text{old}) + \Delta w_{ij} \tag{18.15}$$

Step 10. Check stopping creation which may be maximum number of epochs or when error becomes zero (calculated output becomes equal to desired output).

In the same way solution of load prediction can be trained by time-controlled recurrent method (TCR).

18.3 APPLICATION OF IOTS IN ELECTRIC GRID

Nowadays, the IOT plays a critical role in the smart grid. Smart grid is partly enabled by the IOTs, as its technological and infrastructural components are largely IOT based. Sensor-enabled IOT devices, appliances, and hubs that govern a smart house or any other connected environment provide statistics on energy consumption. This information is then utilized to assess electricity usage, compute costs, control appliances remotely, make load distribution decisions, and discover faults. In the modern era, it is necessary to apply proper data analysis and strategic management of for effective utilization of the smart grid. Sensor technology plays a critical role because it is these components that allow consumers to track their energy consumption. Sensors in smart appliances create and report status data on a continual basis, allowing for monitoring and control. Smart meters collect data on energy usage and display a complete picture of energy consumption in the home, including loads and cost estimates. Figure 18.7 shows the path of strategic management of IOT-based smart grid.

Smart grid data management is challenging because to the type, dispersion, and real-time constraints of individual needs. Strategic management and data analysis methodologies are ideal for this type of application for improved and effective data management. Utilities will be able to do things they've never been able to do before, such as better understanding consumer behaviour, conservation, consumption, and demand, tracking downtime and power outages, and so on, thanks to the massive amount of data. Those utilities that lack the systems and data analysis expertise to deal with these data, on the other hand, will confront challenges. The management of an organization's resources in order to attain its goals and objectives is known as strategic management. Strategic management includes setting objectives, analysing the competitive environment, analysing

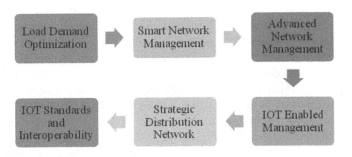

Figure 18.7 Path of strategic management of IOT-based smart grid.

the internal organisation, evaluating strategies, and ensuring that management executes the plans across the organisation.

18.3.1 Design of IoT-Based Smart Grid

The system, identified as "Solar-Powered Smart Grid" is a grid-connected system type and is connected to the main grid with a "Three phase in low voltage" type connection. Its rate is 3.45 kW and annual energy production of 4947.32 kWh resulting from 27 modules, with a surface of 47.66 m^2 and consist of 1 generator. This system includes 3.45 kW of the solar panel has been sourced as an appropriate supplier being able to provide 3.45 kW worth of power in a 14 solar panel array. These panels have a 10-year warranty and an 80% power output guarantee for 20 years, ensuring the panels will not have to be replaced during the lifetime operation of the system. In this project a life cycle assessment of the electricity generated with a solar-powered smart grid using amorphous silicon module produced by Flexcell. The Flexcell module is evaluated per square meter. PV electricity is analyzed per kWh of Busbar. In compliance with the ecoinvent quality guidelines, life cycle inventories of 3.45 kW PV plants on a flat roof with Flexcell module are set up. The carbon footprint of Flexcell module is 19 kg. CO_2-eq. Per square meter. Major contribution stems from the emission of chlorofluorocarbons in the ETFE feedstock (52%), the CO_2 emission from the combustion of natural gas (23%) and the disposal of the module in a municipal waste incineration (10%). Components and materials used in the manufacture of a 4.88 m^2 Flexcell PV module, including production losses. In the modelling of a solar charging station, create a framework based on a specific parameter that is used to generate electricity and meet client demand. A solar charging station simulation is a model constructed with specific boundary conditions that approximates the behaviour of a solar charging station. The solar charging station model is a well-defined description of the simulated parameter that includes crucial aspects including technical, administrative, functional, and physical properties. In the recent scenario, simulation through data analysis is a vital process, and we model the system according to specific parameters, where data follows big data qualities such as volume, velocity, and variety. However, when a choice affecting the design of a solar charging station or the generation of electrical energy for any solar charging station with a kW capacity is made based on such simulation, it is critical to guarantee that no concealed defect invalidates the model or the simulation's conclusion. A pre-feasibility study of a solar charging station is usually conducted prior to installation and operation. Tables 18.1 and 18.2 shows the data of technical parameters for solar-powered smart grid system and energy storage system, respectively. Figure 18.8 shows the architecture of solar-powered charging station, which is composed of three charging points CP1, CP2, and CP3.

Table 18.1 Technical data of PV system

Number of PV Panel	10
Number of Inverter	1
Total surface area required for modules	47.66 m²
Phase 1-Power	1.15 kW
Phase 2-Power	1.15 kW
Phase 3-Power	1.15 kW
Total Power	3.45 kW
Energy/KW	389.37 kWh
Balance of System	75%
Total annual energy	4947.32 kWh
String per module	2×5 = 10
Total Accumulation Capacity	2.90 kWh
AC Junction Box	1
DC Junction Box	1
DC Cable	20 m
AC Cable	20 m
Average Generation	12 Unit/Day

Table 18.2 Technical data of energy storage system

Number of batteries per string	4
Number of strings	2
State of charge minimum	50%
Depth of discharge maximum	50%

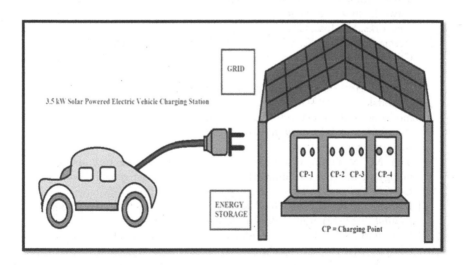

Figure 18.8 Architecture of solar-powered charging station.

18.4 CONCLUSION

The current energy distribution system is gradually evolving into a very big and complicated structure in which information and communication technology is playing an increasingly significant role. The term "smart grid" is now commonly used to describe it. Within the smart grid, there are virtually an infinite number of conceivable applications for ICT subsystems, and this is not to argue that the smart grid is a fixed structure with fixed capabilities. AI, particularly artificial neural networks, is a unique offering of ICT to the smart grid. Here we present our efforts to help the smart grid evolve into a more advanced, dependable, and secure system.

REFERENCES

[1] Dimitrijević, M., Milojković, J., Slobodan Bojanić, S., Nieto-Taladriz, O., & Litovski, V. (2013). ICT and power: new challenges and solutions. *International Journal of Reasoning-Based Intelligent Systems*, 5(1), 32–41.

[2] Kalogirou, S., Metaxiotis, K., & Mellit, A. (2010). Artificial intelligence techniques for modern energy applications. In *Intelligent Information Systems and Knowledge Management for Energy: Applications for Decision Support, Usage, and Environmental Protection* (pp. 1–39). IGI Global.

[3] Gungor, V. C., Sahin, D., Kocak, T., Ergut, S., Buccella, C., Cecati, C., & Hancke, G. P. (2011). Smart grid technologies: communication technologies and standards. *IEEE Transactions on Industrial Informatics*, 7, 529–539.

[4] Zhang, N. (2009). A survey of the development and application of artificial intelligence technology. *Coal Mine Machinery*, 2(02), 4–7.

[5] Riley, D., & Venayagamoorthy, G. K. (July 31–Aug. 5, 2011). Characterization and modeling of a grid connected photovoltaic system using a recurrent neural network. In *Proceedings of IEEE International Joint Conference of Neural Networks*, San Jose, CA.

[6] Verma, P., Sanyal, K., Srinivasan, D., Swarup, K., & Mehta, R. (May 2018). Computational intelligence techniques in smart grid planning and operation: A survey. In *Proceedings of the 2018 IEEE Innovative Smart Grid Technologies – Asia (ISGT Asia)*, Singapore (vol. 22–25, pp. 891–896).

[7] Zografski, Z. (1991). A novel machine learning algorithm and its use in modeling and simulation of dynamical systems. In *Proceedings of Second Annual European Computer Conference, IEEE COMPEURO'91*, Bologna, Italy (pp. 860–864).

[8] Yun, M., & Yuxin, B. (2010). Research on the architecture and key technology of Internet of Things (IoT) applied on smart grid. In *International Conference on Advances in Energy Engineering* (pp. 69–72).

[9] Kaur, M., & Kalra, S. (2016). A review on IOT-based smart grid. *International Journal of Energy Information and Communications*, 7(3), 11–22.

[10] Viswanath, S. K., Yuen, C., Tushar, W., Li, W. T., Wen, C. K., Hu, K., et al. (Oct. 2016). System design of the internet of things for residential smart grid. *IEEE Wireless Communications*, 23, 871–880.

Chapter 19

An Analysis of Artificial Intelligence Initiatives and Programmes in India

Ishfaq Majid and Dr. Y. Vijaya Lakshmi

CONTENTS

19.1 INTRODUCTION

Education 4.0 requires AI integration in the learner-centred education system. The recent years have seen rapidly increasing changes in learning pedagogy significantly facilitated by information and communication technology (ICT) [1,2]. AI and deep learning (DL) are creating a massive shift in the area of technology and among the lives of citizens across the world. AI is likely to transform the way we live and work. Its adoption has been described as the fourth industrial revolution due to its enormous potential [3]. India is continuously working toward the new technological innovations and thereby the funds are being allocated for implementing the project of AI. In a study by Brookings Institution, India is among the top 10 nations in the world in technological advancements and funding for AI. In 2017, the "Union Ministry of Commerce and Industry" set up Artificial Intelligence Task Force. It was established with a purpose of "embedding AI in Economic, Political, and Legal thought processes so that there is the systemic capability to support India's ambition of becoming one of the leaders of AI-rich economies." AI possess an important role in India's economy, including agriculture, industry, and various other sectors which include finance, transportation, public administration, and defence. As a result, it has a significant impact on the GDP growth of the country [4]. According to a research carried out by the "Center for Security and Emerging Technology (CSET)" in March 2021, India is the fourth-largest generator of AI-related scholarly articles and is amongst the top ten AI patent-producing countries. There are some challenges for bringing Artificial Intelligence to India, it includes lack of proper infrastructure, lack of awareness for innovations and digital divide, etc. But the Government is taking various initiatives to overcome these barriers and challenges and to move gradually in the race towards AI. It is believed that India's AI market is likely to reach $7.8 billion by 2025 [5].

19.2 ARTIFICIAL INTELLIGENCE (AI)

AI lays the foundation for various other concepts such as data transmission, software creation, and computing. It aims at transforming the information and communication technology (ICT) sector by various advancements of technology like "deep learning, machine learning and natural language processing" [6]. It can be understood as a field of computing where the focus is primarily on the transmission of anthropomorphic intelligence and thinking into machines that can help and support humans in many aspects. Aizawa [7] defines AI as the study of engineering and science where the development of intelligent machines and various other computer programs are being undertaken. Alshaikhi and Khayyat [8] define AI as the capability of a machine in imitating the intelligent human behaviour or an individual's ability for achieving goals in various situations. The term was first used by John McCarthy in 1956. In 1970, the first international conference on AI was held in Washington. AI is gradually emerging and becoming stronger in various areas such as technology, mathematics, engineering, and physics. This has resulted in a massive shift in all these fields which the current generation is witnessing now [9]. AI is believed to be a functional future in the scientific and technical world where people communicate, learn, and share ideas and opinions with the assistance of soft and hard technologies. The two commonly discussed AI capabilities include (a) tasks that are performed repetitively and automatically by analysing and consequently predicting the outcome/s based on data labelled by humans, and (b) improving human decision abilities by posing problems for which answers are sought through algorithm/s developed by human beings [10]. AI-based technologies have an impact on all sectors including the higher education sector. These technologies are gaining popularity in higher education for improving the quality of learning. Higher education has seen an increased necessity for utilizing the latest technologies to provide online instruction since the pandemic began in 2020. Students can benefit from AI's ability to automate and democratize individualized adaptive learning. It will aid in the reduction of learning gaps and the generation of learning interests among students, as well as the improvement of learning capacity, language affinity, and learning pace. Though there is very little understanding of AI, but it is a technology that is revolutionizing every aspect of life. It allows people to rethink how information is integrated, analysed, and used in the insights to make better decisions. In short, AI is becoming an increasingly important part of our daily life by allowing us to do more complex and time-consuming work with the touch of a button.

19.3 ARTIFICIAL INTELLIGENCE IN INDIA

The NEP 2020 recognizes the significance of AI and lays emphasis on the significance of training each individual for an AI-driven economy. Over the next few decades, AI is going to change global competitiveness, giving a strategic edge to early adopters. The Indian government is not lagging behind in AI initiatives. In the year 2021, the US-India AI flagship was started. The objective of the program was to provide stakeholders from the United States and India an opportunity to share their ideas and experiences for fostering AI innovations. In May 2021, the Corporate Affairs Ministry (MCA) launched an updated version of their portal MCA 21 version 3.0. The new update will utilize the latest technologies such as ML, AI, and data analytics. Under NEP 2020, the Government of India constituted the autonomous body "National Research Foundation (NRF)" for giving a boost to research across various segments. Under NEP 2020, AI will now be a part of Indian school curriculum. "National Council of Education Research & Training (NCERT)" has taken steps toward developing a new curriculum framework for Indian schools. Under this new curriculum, AI will be introduced at the secondary level [11]. The GOI has taken many other initiatives in the area of AI, there all initiatives, documents and papers are elaborated in the following section.

19.4 NATIONAL STRATEGY ON AI 2018

The discussion paper on "National Strategy for Artificial Intelligence" was published by NITI Aayog in June 2018. This strategy plan demonstrates that India possesses the power and attributes necessary to place itself among the global AI leaders. It also looks at how India might use transformational technology to promote social and inclusive growth in keeping with the government's development philosophy. It is believed that AI has the capability to add significant value to a wide number of industries and it is rightfully described as a transformative technology.

The NITI Aayog decided to focus on five sectors (healthcare, agriculture, smart cities & Infrastructure, education, smart mobility, and transportation) for taking maximum benefit from AI. In addition to mentioning the benefits, the paper also discusses about the barriers that should be addressed at the earliest for achieving the AI goals. The barriers mentioned in the paper are "lack of knowledge in AI and its lack of supportive data ecosystems, high resource cost and lack of proper awareness for adopting AI, privacy and security concerns and lack of collaborative approach for adopting and use of AI." The paper believes that the research of AI in India is in the infancy stage, and it requires collaborative efforts and large-scale interventions for scaling it up. It proposes a two-tier structure or method for addressing aspirations of India's AI research. This two-tier structure or method includes "Centre of Research Excellence (CORE)" and "International Centres of Transformational AI (ICTAI)." The main focus of CORE is to develop a proper understanding of the available research in AI and pushing technology towards creating new knowledge as shown in Figure 19.1. The ICTAI focus will be on the development and employment of application-based research. The document also focuses on the adoption of AI in various startups, the private sector, PSUs, and government entities. It believes that by adopting AI in all these sectors, it will create a good cycle of demand and supply. The document also gives its suggestions on very important aspects such as data processing and ensuring privacy and security. For this purpose, the paper recommends the establishment of a framework for data protection and sectorial regulations.

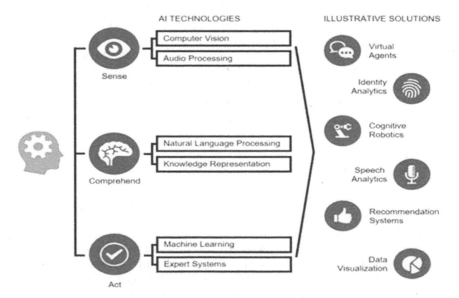

Figure 19.1 Emerging technology.

19.5 AI FOR ALL INITIATIVES

The Ministry of Education (MOE), Government of India launched "AI for All" in cooperation with the Intel and CBSE in August 2021. The main aim of this initiative was to create an understanding of AI among Indian citizens. The "AI for All" is a self-learning online program that aims at creating awareness about AI. The self-learning program is developed for all individuals like students, professionals, and senior citizens. The program wants to attract those users who want to have a "Digital First Mindset." The "AI for All" is divided into various sections. The first section is related to AI Aware which talks about basic understanding of AI, Human Intelligence. It also helps the learners to understand the difference between AI and non-AI machines. Section 19.2 of the program is related to AI appreciate which describes various domains of AI and their impact on various industries. It also elaborates the various principles of AI ethics and responsible AI. This program is accessible in almost 11 Indian languages and is of 4-hour long duration. After completing both sections, the learners will be offered digital badges that can be shared through social media platforms.

19.6 AI GAMECHANGERS PROGRAM

National Association of Software and Services Companies (NASSCOM) in collaboration with Microsoft launched the "AI Gamechangers" program in the year 2021. The program aimed at promoting AI-led innovations in India. This will lead to speed up the adoption of AI in India. The program will serve as a platform for startups, academia, enterprises, government, and NGOs to showcase their AI-based products and solutions. It will provide reach and impetus to drive AI innovations in India. The AI innovators success stories will be featured in the annual compendium. It will be released at the Xperience AI Summit. The program will help India in becoming the larger ecosystem to leverage the opportunity and become a global powerhouse in AI-led innovation. The entries will be judged on three major areas of problem selection, solution invention, and impact achieved by a renowned panel of industry leaders and subject matter experts. The nomination period was kept open from March 4 through April 16, 2021, with screening and final presentations to the Steering Committee and expert panel following.

19.7 AI ACADEMIC INSTITUTIONS AND CENTRES IN INDIA

The Centre for Excellence in AI (CAI) was setup in April 2018 in IIT Kharagpur. It was established due to the increasing importance of AI in India. The centre is being funded by Capillary Technologies. The focus of the centre is on the four quadrants: Entrepreneurship, Teaching, Research, and Industry Projects. The Centre for AI and Robotics (CAIR) is associated with DRDO and is considered as a premier laboratory. The laboratory is situated in Bangalore, Karnataka, and was founded in 1986. The laboratory is mainly involved in carrying out research and development in several fields of AI, Robotics, Information and Communication Security and Command & Control. The Robert Bosch Centre for Data Science and AI (RBC-DSAI) was established at IIT Madras and was established in the year 2017. The centre was established to undertake research in AI and Data Science. It includes "machine learning, deep learning, network analytics, reinforcement learning." The centre is regarded as the official data science and AI partner to the "Ministry of Statistics & Programme implementation." The centre is continuously participating in various initiatives by various industry bodies such as NASSCOM and CII. The AI Group (AI@IISc) is a group of students and faculties at IISC Bangalore. These students and faculties are sharing their interest in the area of AI, machine learning, and data science. The centre offers short-term courses including bachelor's and master's courses in areas such as AI, deep learning, machine learning, fintech, image processing, reinforcement learning, and computer vision. The Department of AI at IIT Hyderabad was established in the year 2019. It provides the students with a holistic knowledge of AI and makes them leaders in AI industry. The IIT Hyderabad now offers a BTech program in AI which makes it the first institution in India to do so and globally the 3rd institution.

19.8 STATE OF AI IN INDIA 2021

For understanding the AI development in India, Analytics India Magazine (AIM) in collaboration with TAPMI developed the State of AI in India study. The study comprises eight sections. Section 19.1 of the study provides a summary of the AI market in India. The study reveals that as of July to August 2021, the Indian AI market is valued at $7.8 Bn. This indicates a 22% increase in the market over 2020. There are at least 10,900 AI personnel working in India. The AI professionals in Mumbai have the highest median salary at around 17.3 lakhs. The AI professionals in Bengaluru and Delhi have median salaries of 16 lakh and 13.9 lakh, respectively. This is being followed by Pune and Hyderabad where the median salary of AI professionals is 13.2 lakhs and 12.6 lakhs, respectively. In Chennai, the salary is around 10.8 lakhs. There are around 14,500 positions available in India as of July to August 2021. Section 19.2 of the study deals with AI Market Size by Type of Company. On the basis of the type of company, the MNC IT, Technology and Electronics categories are possessing the highest market share of 38.8%. The overall market value of AI services provided by specialized AI enterprises is $169.6 million. As far as the salary of AI personnel is concerned, at least 23.5% of personnel are in between 3 and 6 lakhs salary level. Around 17.5% of professionals are in 6–10 lakhs salary level and 15.6% of professionals are in 15–25 lakhs salary level. The overall AI market is predicted to grow up at a CAGR of 27.5% from 2022 to 2026, valuing the market at $23 billion by 2026. Similarly, the AI domain, excluding IT services, is predicted to grow by 22.5% in 2025, valuing the AI market (without IT) at $12.3 billion. Bengaluru is among the cities offering the highest jobs in AI and standing at 29.9%. The second is Delhi which is advertising the AI jobs at a percentage of 17.9.

19.9 75 & 75 INDIA'S AI JOURNEY

AI has emerged as a cutting-edge technology that is reshaping the economy's present and future. To honour these technological advancements and milestones, MeitY, NeGD, and NASSCOM have teamed up to highlight the most intriguing AI use cases from a variety of industries, including spanning healthcare, agriculture, education, governance, and financial services and create a resource for the readers. To honour 75 years of independence, the compendium features 75 use cases that have positively and successfully benefited the people and governance across the country. For the comfort of readers, these 75 cases have been categorized under four heads which include "government, startups, academia, and enterprises" to fathom the progress of usage driven by such technologies.

During COVID-19, the Central Government created MyGov Corona Helpdesk which was world's largest WhatsApp chatbot for handling the queries of people in English and Hindi. The chatbot was launched in March 2020 and till today it has served to at least 60.7 million individuals with 240 million messages. The sensor-based Internet of Things (IoT) tools were introduced by the Union Jal Shakti Ministry to monitor the implementation of regular tap water schemes in every household. The states such as Sikkim, Manipur, Goa, Maharashtra, and Uttarakhand have started to roll out this technology and the various other states such as Gujarat, Bihar, Haryana, and Arunachal Pradesh have issued tenders for implementing the scheme. To screen people for cataract, the Tamil Nadu e-Governance (TNeGA) developed an AI-based mobile app "ePaarwai." The app is being tested in various other districts of Tamil Nadu. The Government of India's Digital India Corporation (MyGov) developed MyGov Saathi during the COVID-19 pandemic. It was an AI agent which addressed the FAQ of users from a variety of available options. The AI agent can handle 300,000 users a day and around 20,000 users per minute. The Government of Telangana implemented the Real-time Digital Authentication of Identity (RTDAI) to authenticate pensioners. The AI tool can identify the old pictures in the database from the latest photographs. The AI tool can also find the mismatch in the spellings of names. The tool has an accuracy of 93% which keeps on increasing. In a similar manner, the Government of Andhra Pradesh introduced the AI-based Real-time Beneficiary Identification System (RBIS) in the year 2020. The AI tools come with face recognition on a real-time basis which helps the Government in distributing the pension to at least 61 lakh pensioners in a transparent manner. The Government of Uttar Pradesh has introduced and implemented the video analytics platform JARVIS in at least 70 jails across the state since 2019. This tool helps the Government authorities in monitoring the jails through CCTV surveillance and flag any unlawful activity that takes place inside the jail. The Government of Telangana has introduced the BVLOS drone flights in the state to deliver vaccines, blood and medical samples, and organs. These deliveries are taking place in 16 Primary Health Centres in Vikarabad area. The DRDO has developed chest X-ray and AI for detecting the COVID-19. Through the 5G network, it has been possible to provide access to radiologists anywhere through its public cloud and AI-based infrastructure. Presently, the Government has signed an MOU with Government of Karnataka for deploying ATMAN across 194 PHCs. For making digitization even more powerful, the Government of India created an online portal SUPACE for high courts of India and the judges of these courts have been asked to use it for enhancing their efficiency. The state of Tamil Nadu has developed a cost-effective DL-FRS a deep learning approach which uses face recognition to mark the attendance of students when they enter school. This AI bases technique helps the school authorities in easy tracking of attendance records. The state of Maharashtra in collaboration with Haptik Infotech developed an AI-enabled chatbot that can provide the information on Public services. The chatbot is comfortable in Gujarati and Marathi languages as well. In the state of Uttar Pradesh, the Police is using AI in catching criminals in the state. The state has developed an AI-based app "Trinetra" which contains a record of 5 lakh criminals which comprises of photo, address, and criminal history of each criminal. This AI-based app not only finds criminals but also helps in finding their associates as well. The National Highway Authority of India has introduced AI-based face recognition for monitoring the attendance of its field staff posted at different locations (Table 19.1).

Besides this, there are several other AI startups in India such as Artivatic, Verloop.io, Unbxd AI, and Agri10X. As far as higher education is concerned, AI is being used in various higher education institutions such as IIIT Hyderabad, IIT Kharagpur, IISC, IIIT Delhi, IIT Madras, IIIT Delhi, IIT Bombay, IIT Ropar, and ISI Kolkata.

Table 19.1 List of AI startups in India

Name of the startup	Brief description
AgNext	It is an AI-powered application used to identify the quality of agricultural produce.
NIRAMAI	It is a Bangalore-based app which utilizes machine intelligence to detect breast cancer among patients.
TRICOG	This AI-powered startup has the ability to provide accurate ECG reports within a few minutes of taking ECG at remote centres. It was founded in 2015.
HAPTIK	It is an online supplement store where a variety of health and wellness products are available. It is basically an Intelligent Virtual Assistant to provide diet and fitness advice seamlessly.
JAGADISH K. MAHENDRAN	It is an AI-powered voice-activated to help visually impaired persons to navigate and perceive the world around them.
COGNIABLE	It is an AI-powered app which screens the uploaded video of any child having Autism. The platform provides inputs from 07 behavioural psychologists and 04 technical experts and hence provides intervention plans for these children.
SPEKTACOM	It is a kind of sensor technology used on the cricket bats for convergence of data with insights from cloud-based data analytics, AI and machine learning.
LEARNING MATTERS	It is a voice assistant for improving the English proficiency of teachers.
CHALO APP	Its old name was Zophop and was launched in 2014. It is a kind of trip planner app that can provide various suggestions to users for the best mode of public transport.
TAGHIVE	It is a multi-stakeholder solution which empowers teachers, students, parents, and administrators. It automates various tasks like attendance taking, question framing and report making. Overall, it improves the productivity of teachers.
SHOPTIMIZE	It is an AI-Driven platform for boosting the revenues of online stores. This AI-powered platform helps the consumer brand for achieving consistent revenue growth.
AQUACONNECT	It is an AI-powered platform for providing data-driven shrimp framing advisory services. It is helpful in finding diseases and advises the farmers to take corrective measures.
CROPIN	It is a climate-smart advisory-based platform which helps in scheduling and monitoring various farm activities. In addition, it provides a seven-day weather forecast derived from various observation and forecast models.
ARYA. AI	It is an AI and machine learning-powered platform being used in financial services. It is being used to automate the functions of the insurance sector.
CULTYVATE	The platform is being used to calculate the soil moisture for deciding the optimum water management. The platform's advanced sensors and technologies keep a track of the usage of water. The framers with the help of this platform were able to check the ground-level data on their smartphones.

(Continued)

Table 19.1 (Continued)

Name of the startup	Brief description
DECIMAL TECHNOLOGIES	It is an AI-driven lead management solution for implementing an automated, end-to-end lead management solution to create, process and track sales lead from generation to conversion for IndusInd Bank.
VPHASE	This platform is being used by HDFC bank for having an eye on sales. It also provides performance reports of all the branches and digital channels.
LOCUS AI	LOCUS AI is an AI-enabled urban ladder for improving efficiency and productivity. It is being used to automate logistics operations and optimize last-mile delivery costs.
TARTAN SENSE	It is a Bengaluru-based agritech developed for carrying out weeding and pest control. The Robot is trained with images of crops and weeds so that it can differentiate between them.

19.10 RESPONSIBLE AI #AIFORALL

In 2021, NITI Aayog published the Responsible AI #AIFORALL document in cooperation with World Economic Forum Centre. The document aims at establishing broad ethical principles for designing, developing, and deploying AI in India. The paper follows expert consolations and interviews with key agencies adopting AI solutions in India. These consultations have been divided into the system and societal considerations. The system considerations are generally concerned with overall principles. The societal consideration deals with the impact of automation on job creation and employment. The article also provides a look into the legal and regulatory approaches to AI system management. Accepting that there are multiple stakeholders in the current AI ecosystem, including private, public, research institutes, legal bodies, and so on, it emphasizes the importance of laying the groundwork for common acceptable behaviour among them, as well as immediate clarification in the event of a disagreement. There is also a discussion of guidelines for responsible AI system administration, which are formed on the basis of the country's approved value system and are in line with worldwide norms. Equality, safety & reliability, inclusivity & non-discrimination, openness, accountability, and privacy & security are some of the fundamental principles that have been investigated.

19.11 CONCLUSION

AI has always been a kind of technology which is attracting all the people around the world. It is considered the most transformational technology available today. Governments around the world are increasing their investment in AI and are figuring out the various ways for applying and encouraging the application of it. Over the next few years, AI is expected to change global competitiveness, giving a strategic edge to early adopters. The initiatives, documents, and papers discussed in this study provide insights into the development of AI in India. The GOI has started their various initiatives for employing the AI policies for taking maximum benefit from AI and it becomes their responsibility to speed up their initiatives towards AI for the betterment of the nation. The IT sector of India has always been a centre of great minds and it has played a vital role in the economic development of the nation. The government till date has started various AI-powered initiatives such as biometric identification, face recognition, etc., and it is expected that more such technologies will be within India very soon.

REFERENCES

[1] Sinha, E., & Bagarukayo, K. (2019). Online education in emerging knowledge economies: exploring factors of motivation, de-motivation, and potential facilitators; and studying the effects of demographic variables. *International Journal of Education and Development using Information and Communication Technology, 15* (2), 5–30.

[2] Tijani, O. K., Obielodan, O., & Akingbemisilu, A. (2020). Usable educational software: teacher educators' opinion about Ọpọ́n-Ìmọ̀ technology enhanced learning system, Nigeria. *International Journal of Education and Development using Information and Communication Technology, 16*(1), 88–106.

[3] Srivastava, S. K. (2018). Artificial intelligence: way forward for India. *Journal of Information Systems and Technology Management, 15*, 1–10.

[4] Dhanabalan, T., & Sathish, A. (2018). Transforming Indian industries through artificial intelligence and robotics in industry 4.0. *International Journal of Mechanical Engineering and Technology, 9*(10), 835–845.

[5] Mayer, V. K. (2022). How far AI has come; what the future looks like. *Business World*. Retrieved from https://www.businessworld.in/article/How-Far-AI-Has-Come-What-The-Future-Looks-Like-/03-01-2022-416525/

[6] Haldorai, A., Murugan, S., & Ramu, A. (2020). Evolution, challenges, and application of intelligent ICT education: an overview. *Computer Applications in Engineering Education, 29*, 562–571.

[7] Aizawa, K. (1992). Connectionism and artificial intelligence: history and philosophical interpretation. *Journal of Experimental & Theoretical Artificial Intelligence, 4*(4), 295–313.

[8] Alshaikhi, A., & Khayyat, M. (2021). An investigation into the impact of artificial intelligence on the future of project management. *Paper presented in International Conference of Women in Data Science at Taif University*, 1–4. IEEE.

[9] Benko, A., & Lányi, C. S. (2009). History of artificial intelligence. In *Encyclopedia of Information Science and Technology*, 2nd edition (pp. 1759–1762). IGI Global.

[10] Alam, A. (2020). Possibilities and challenges of compounding artificial intelligence in India's educational landscape. *International Journal of Advanced Science and Technology*, 29(5), 5077–5094.

[11] Gazala, N. (2021). AI in India: initiatives from the Indian Government in 2021. *India AI*. Retrieved from https://indiaai.gov.in/article/ai-in-india-initiatives-from-the-indian- government-in-2021

A Vital Fusion of Internet of Medical Things and Blockchain to Transform Data Privacy and Security

Sumit Kumar Rana, Arun Kumar Rana, and Sachin Dhawan

CONTENTS

20.1 INTRODUCTION

As we are in the era of industry 4.0, the use of different technologies such as machine learning, artificial intelligence, Internet of Things, etc., has increased to automate the conventional process of industries. When the

DOI: 10.1201/9781003355960-20

above-mentioned technologies were incorporated with the industrial process, different positive outcomes were seen like lesser human intervention, reduced error, reduced delays in data exchanges between processes and devices etc. After noticing these benefits, researchers worked on the expansion of using these technologies in different domains such as residential infrastructure, financial domain, healthcare sector, etc. When these technologies were incorporated in healthcare, a new term healthcare 4.0 came into existence [1]. This revolution brought a new way of execution of medical processes and data management. It brings the Internet of Medical Things (IoMT) to the system [2]. Different components of the IoMT system are shown in Figure 20.1. It consists of sensors that provide data from various sources. It has a network component that is used to exchange information between different entities. It has some tools for data storage and applications for accessing the data. Nowadays electronic information systems are now being utilized in medical care as a result of the fast development of information technology, and a great quantity of medical data is being generated every minute such as lab reports, medication details, diagnostic reports, etc. The appropriate use of this medical data can not only aid in the early detection of infectious illnesses and the preparation of protective measures but same can be used to resolve any medical disputes [3]. As a result, it's important to research how to use medical data in an effective manner.

Patient's treatment situations may be recorded quickly and accurately using electronic medical data and treatment experiences can be shared in a network of different medical entities. However, privacy attributes will be in danger if the shared medical information is fraudulently abused. As a result, regulating medical data access is a pressing concern. Medical institutions outsource the maintenance of encoded health data to a third party that reduces computing costs and saves local storage space but also allows for faster data retrieval. Furthermore, having entire EHRs in the cloud allows

Figure 20.1 Components of IoMT architecture.

healthcare practitioners to keep track of their patient's health and give appropriate medical services throughout diagnosis and treatment [4].

The cloud server is centralized, if one cloud model fails, the entire cloud server will become unavailable [5]. The healthcare business has seen substantial changes in e-health operations as a result of the advent of novel technologies such as IoMT. To process and analysis of the medical data, received from IoMT equipment, clouds are being used. Because of this, patients can be diagnosed by the health professionals in a faster and more accurate manner. However, this approach also increases the fear of exploitation of patients' health data [6]. The privacy of patients' health data must be safeguarded. However, several examples of large-scale security breaches of cloud systems created panic about the protection of health data. As the use of the Internet and its related online technologies increased, the number of cases of cyber attacks also increased [7]. Most of organization has been attacked by at least one cyber attack. A list of recent attacks is shown in Table 20.1. The data are taken from CyberEdge Group 2021 Cyberthreat Defence Report.

Figure 20.2 shows the graph of cyber attacks from 2014 to 2021. This information is taken from CyberEdge Group 2021 Cyberthreat Defence Report. We can analyze that from 2014 to 2021 percentage of firms suffering from cyber attack has increased.

Indeed, hostile attackers are increasingly targeting servers where health data is hosted, as personal health data is quite valuable in the black market. As a result, health professionals must implement more stringent security measures, raising the costs of maintenance for these services. Countermeasures for these breaches and after-breach procedures should be

Table 20.1 Details of some recent cyber attacks

Year of cyber attack	Description of cyber attack	Location
March 2022	DDoS attack on national telecommunications authority and disrupted internet services for a week.	Marshall Islands
March 2022	Espionage attacks to find secret information about military websites.	India
March 2022	Hackers displayed anti-government and anti-invasion images and messages on government websites.	Russia
March 2022	Hackers installed backdoors and deployed malware in various organizations related to the energy sector.	USA
March 2022	The network of the national research council, a state-funded research agency, was penetrated.	Canada
March 2022	DDoS attacks on telecommunication providers to disrupt the online services of government websites.	Israel
February 2022	DDoS attack on defence ministry websites and largest bank of the country to disrupt the services.	Ukraine

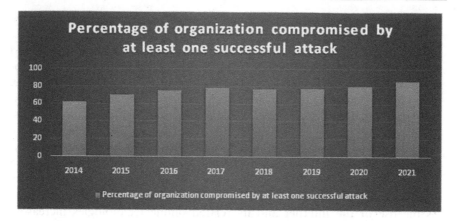

Figure 20.2 Percentage of at least one successful graph on industries.

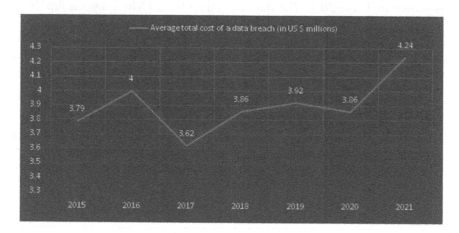

Figure 20.3 Average total cost of a data breach (in US $ millions).

inspected delicately. These data breaches can cost very high to an organization. The average total cost of a data breach for the previous seven years is shown in Figure 20.3. (source: IBM's data breach report 2021). So, an efficient access control solution is required to avoid these data breaches.

The average cost of a data breach varies as per the country or region as shown in Figure 20.4. This cost is highest in the United States and lowest in India. We can analyze from the graph that this cost has increased from the year 2020 to 2021.

The average cost of data breach varies as per industry as shown in Figure 20.5. As compared to 2020, cost of breach has increased in 2021 in different industries. This figure shows that healthcare is suffering from the highest cost of data breach. So, we need a solution that can help in preventing this data breach in the healthcare industry.

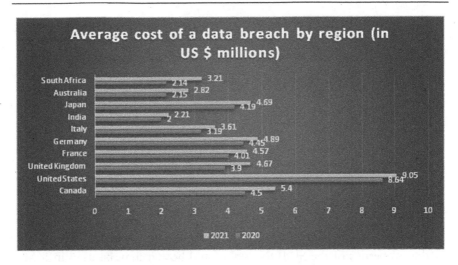

Figure 20.4 Average cost of a data breach by region (in US $ millions).

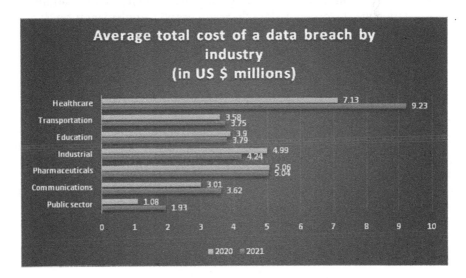

Figure 20.5 Average cost of a data breach by industry (in US $ millions).

From the emerging technologies, blockchain technology can be the perfect fit to solve the security and privacy problem of the healthcare industry. With its decentralized character, this technology has shown massive potential in a variety of services provided by medical industries. As a result, a surge in using this technology has been noticed in recent times [8]. Access control systems can effectively prevent unauthorized access to EHR resources while allowing authorized organizations to retrieve data quickly [9]. Our suggested solution is to build an access control framework supported by blockchain technology.

The chapter is structured as follows. In Section 20.2, a literature survey of IoMT is performed. In Section 20.3, issues in healthcare 4.0 is discussed. In Section 20.4, security and privacy requirements for IoMT are described. In Section 20.5, challenges related to IoMT security and privacy are discussed. The role of blockchain technology in IoMT is explained in Section 20.6, and the conclusion and future research directions are presented in Section 20.7.

20.2 RELATED WORK

20.2.1 Internet of Medical Things

For IoMT-enabled State Health Society (SHS) applications, the authors presented a lightweight cryptographic method. The contribution looked at the properties of the SIMON cypher and how it may be used in IoMT-enabled SHS applications to improve performance. To reduce the computational burden imposed by encryption, the system advised adding a modification via the original SIMON cryptographic implementation. It also allowed for the realistic balance of performance and security to be maintained. The system, however, did not produce satisfactory results [10].

In comparison to asymmetric encryption approaches, the results of the proposed technique decreased the computing cost. In addition, when compared to alternative lightweight techniques, the system had a lower security risk [11].

The findings of the security study indicated that the system can safely and effectively transfer data in IoMT-enabled SHSs while also being computationally efficient [12].

The authors proposed a method for securing IoMT. The system was protected against a variety of threats. In addition, the system performs both official and informal security assessments. Comparison with the relevant methodologies revealed that the system was acceptable and has some enhanced security qualities. However, the system was unable to display the exact security level [13].

20.2.2 Verification and Approval Technique for IoMT

For IoMT, the authors proposed a trivial and strong verification scheme based on the physically unclonable function. This technique does not save any data relating to IoMT devices in server memory. A hybridized oscillator arbiter PUF is used to accomplish system validation. As per the function used for system authentication, the number of keys was around 240. Because the authentication technique is lightweight, it may be used in a variety of designs to enable scalability and resilience. However, the system failed to verify that the client could authenticate the server's communications [14].

These procedures secured the legality of the nodes in the community medical IoT, as well as the communication security. The system, on the other hand, had significant computational expenses and a higher power use [15].

When compared to centralized delegation design, the results of the investigation proved that the suggested architecture provided higher security. The system, on the other hand, was not resistant to prospective assaults [16,17].

In Het-Net, the authors proposed a safe connection for health apps that uses ABE for verification. ABE is used to safeguard health-related data here. This has not only reduced communication overhead, but it has also helped to protect health data from invaders [18]. A high-level protocol specification language is used to implement the complete security method (HLPSL). AVISPA, an automated tool, is used to validate the system codes. However, the system lacked security and was unable to function properly [19].

The authors investigated a lightweight architecture for authentication and permission to support distributed ledger technology-supported IoT networks in medical sector. The authentication method uses random numbers. This allowed the system to create a secure link between IoT devices for data collection. For system validation and assessment, the authors used automated tools and simulators such as AVISPA and Cooja. The technology offered better access control while also providing excellent mutual authenticity. When compared to alternative options, it reduces both communication and computational overhead costs, as evidenced by the experiential outcomes. The system, on the other hand, was inefficient [20].

In the IoMT, the authors proposed an approach which was based on the system's effective matching technique based on the Fishers vector's secondary calculation (FV). In addition, the system made use of different biometric techniques. In addition, the system used a bogus feature in the feature fusion process, which is a common occurrence in the real world. As evidenced by the experience outcomes, the developed framework had a higher recognition rate. It provided increased security as compared to unimodal biometric systems, which are critical for an IoMT platform [21].

20.2.3 Privacy Preserving Approaches for IoMT

The authors proposed an elliptic curve digital signature-based PP technique for IoMT. This solution ensures anonymity in data transported over IoMT to the cloud by using edge computing servers. Despite this, the system's calculation and communication costs were determined to be considerable [22].

The authors suggested a protocol for protecting SHS, in which an invader cannot impersonate a legitimate user in order to get unauthorized access. The experimental analysis was used to analyse the network parameters using the NS3 simulator. When compared to other existing protocols, the gathered findings showed superiority in terms of output, packet delivery, etc. [23].

The authors proposed a method for health data exchange with the help of sensors in multi-party computation. Despite the system's low degree of security, private keys were saved in sensors to communicate with data servers [24].

The authors proposed a user-based secure data transfer system for IoMT. The encrypted copy is generated on the basis of number of users selected. The findings revealed that secure data might be successful when used to provide security in IoT [25].

The system was also assessed to see if it meets the frequency of attribute values, the privacy requirements of hospitals' datasets, and successfully identifies the symptoms. The system did, however, have access to low-security services [26].

Despite the scheme's enormous payload, it was shown to maintain a good degree of perceptual quality. The system provided a significant improvement and was also computationally efficient, allowing it to be used in the IoMT network [27].

20.2.4 Decentralized Access Control

The authors proposed the "MedSBA" approach for storing medical data in SHS, which is both effective and secure. The system's functioning was demonstrated using BAN logic, while the security was established using formal design. Simulating MedSBAs using OPNET software demonstrates the system's efficiency in terms of computing complexity and storage [28].

To demonstrate the system's capabilities against many forms of conceivable attacks, a comprehensive formal security study was undertaken using the widely recognized automated tool, AVISPA. After the analysis, comparative study findings showed that the suggested technique, BAKMPIoMT, outperformed other current methodologies [29].

Furthermore, it avoids medical conflicts because doctor diagnoses and IoT data recorded on blockchain cannot be manipulated with or removed. The system's performance and security evaluations proved that it is suitable for SHS. Insider assaults, on the other hand, are ignored by the system [30].

20.3 SAFETY AND CONFIDENTIALITY NECESSITIES FOR IOMT

Every domain has different types of safety and confidentiality requirements. For IoMT, requirements are more rigorous as compared to IoT. Working methodology is different at each level so the requirements change as per the needs of the particular level of IoMT systems. Level-wise requirement for security and privacy is discussed as follows.

20.3.1 Data Level

20.3.1.1 Confidentiality

Patient health information must be acquired and maintained in line with legal and ethical privacy requirements, which restrict access to only authorized staff. In order to avoid data breaches, appropriate measures must be implemented to protect the privacy of medical data related to particular patients. Data seized by cyber hackers may be sold on illegal markets, placing patients at risk of not just privacy violations but also financial and reputational loss [18].

20.3.1.2 Integrity

The data integrity requirement for IoMT is designed to confirm that information arrives at its intended destination without being interfered with during wireless transmission [31]. Intruders use wireless networks to access unauthorized data which might result in life-threatening situations. Once the data are stored in the system, any illegal amendment in the data should be traced and blocked. Effective procedures for access controls must be implemented and the integrity of information stored on medical servers must be ensured.

20.3.1.3 Availability

Information and related services should be available to the appropriate users when they are needed [32]. So, to prevent these types of situations, healthcare-related applications and services should be accessible at any time instance.

20.3.2 Sensor Level

Medical equipment lags in processing power and suffered from limitations in power which created issues for the security and privacy needs of the network. Some of the capabilities that are desired at the sensor level are discussed in the following sections.

20.3.2.1 Tamper-Proof Hardware

Physical equipment such as sensors can be physically stolen, exposing security information to attackers. Furthermore, attackers can reprogramme the stolen devices and re-deploy them to the system, listening in on conversations without being detected [33].

20.3.2.2 Localization

The two types of sensor localization that researchers are interested in are on-body sensor location and sensor/patient placement. An invasion exposure mechanism that alerts managers to hostile nodes joining the network, is an example of such a measure [34].

20.3.2.3 Self-Healing

IoMT devices should have the capability of self-healing in case of cyber attack. However, because different forms of network assaults need distinct detection and recovery mechanisms, network managers must select which autonomic security strategies should be used [35].

20.3.2.4 OTA Updates

There should be some provision so that any kind of updates can be sent over the air without moving to the location of the devices. Sometimes these OTA updates are misused by the attackers [36]. So, security precautions must be available in case of such attacks. SEDA, for example, is a safe over-the-air updating technique developed for medical systems.

20.3.2.5 Forward and Backward Compatibility

Identification of the faulty sensors and their diagnosis must be performed as fast as possible. If diagnostic signals are transmitted after the occurrence of the fault, then sensors will not be able to decode it. It is known as forward compatibility. Backward compatibility, on the other hand, prevents messages from being read by a sensor that has recently joined the network [37]. Compatibility concerns may be resolved by using OTA programming to distribute the most recent software update as soon as possible.

20.3.3 Medical Server Level

Authorized devices and individuals have the permission to read or update patient data on the medical server level, and data must be protected at all times while stored in databases. Safety and confidentiality problems are increasing rapidly as more paper-based medical information are transformed into electronic medical records (EMRs) [38]. As a result, suitable safety systems are required for IoMT at the server level.

20.3.3.1 Access Control

To avoid unauthorized access, a proper access control is required. Because on each access request, obtaining a patient's permission or consent each time is difficult. Medical servers should be able to change access control restrictions quickly. Policy changes on medical servers might be redundant [39]. An approach is required that reduces overheads in cryptographical computation.

20.3.3.2 Key Management

Key management methods are responsible to build and allocate keys for building secure applications[40]. The second method is key pre-distribution methods which are frequently used to transfer secret keys inside a network before it is completely functioning. These protocols are better suited to sensor networks with low resources since they are simple to construct and do not require a lot of processing.

20.3.3.3 Trust Management

Trust can be defined as the degree to which a node can exchange information with other nodes with safety and trust [41].

20.4 ROLE OF BLOCKCHAIN TECHNOLOGY IN IOMT SECURITY

Because of its significant cryptographic security and immutability, the blockchain has revolutionized the processes of various domains. The major features of blockchain can help to address the difficulty of safe data exchange amongst heterogeneous IoMT devices, ensuring the data's dependability. There are a variety of blockchain platforms that might be a good fit for IoMT devices that rely on centralized cloud networks [42]. The centralization or reliance on an intermediary is the most significant distinction between cloud and blockchain. Furthermore, cloud service providers might circumvent customers' confidentiality and interfere with information without their agreement in specific scenarios [43]. There are multiple IoMT devices that generate the data so ledger size grows rapidly in IoMT, posing a problem for IoMT device's limited capacity to store and handle such a massive volume of data. As a result, IoMT devices might suffer in validating the transactions on the decentralized network [44].

Furthermore, cloud computing may be prone to unlawful data sharing, which compromises users' privacy. Participants at various level are shown

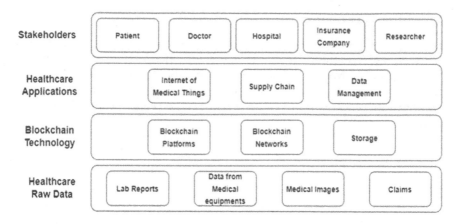

Figure 20.6 Participants at various levels.

in Figure 20.6. Blockchain, on the other hand, has the capacity to empower its users by providing access control via smart contracts, allowing them to restrict access to their data. Before submitting the data to the blockchain network, data is encrypted. Furthermore, the encoded information can be decoded with the matching private key [45–46].

Many alternative blockchain systems, including Ethereum, Hyperledger-Fabric, have been created since Bitcoin to provide diverse utility services other than money. Machine-to-machine micropayment methods are required in modern IoMT systems, and smart contracts are used to protect users' privacy and sensor regulations. The connectivity of different sources with blockchain networks is shown in Figure 20.7.

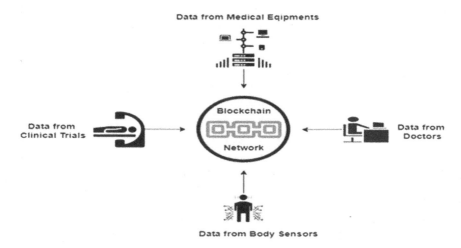

Figure 20.7 Connectivity of different sources with blockchain network.

20.5 CONCLUSION

Fusion of the IoMT with blockchain technology is a very vital solution for the healthcare industry. Various attributes of blockchain technology such as consensus, immutability, transparency, etc., help in preserving the privacy and security of health data. Various papers about IoMT and blockchain technology have been studied. Various issues and problems with IoMT and medical data are identified during the survey. Different security and confidentiality need of IoMT are analyzed. We also discussed that how blockchain technology can help in improving the privacy and security of health-related data. But adoption of blockchain technology is a concern because blockchain technology is in the early stages of development. Existing infrastructure is need to be updated to adopt blockchain technology for various applications.

REFERENCES

[1] Rana, A. K., & Sharma, S. (2019). Enhanced energy-efficient heterogeneous routing protocols in WSNs for IoT application. *IJEAT, 9*(1), 4418–4415.

[2] Lewis, T., Liwen, W., Safa, O., Moayad, A., & Jalel Ben, O. (2020). Blockchain for managing heterogeneous internet of things: a perspective architecture. *IEEE Network, 34*(1), 16–23.

[3] Hassan, W. H. (2019). Current research on internet of things (IoT) security: a survey. *Computer Networks, 148*, 283–294.

[4] Kumar, A., Sharma, S., Goyal, N., Singh, A., Cheng, X., & Singh, P. (2021). Secure and energy-efficient smart building architecture with emerging technology IoT. *Computer Communications, 176*, 207–217.

[5] Zang, W., & Li, Y. (May 2018). Gait-cycle-driven transmission power control scheme for a wireless body area network. *IEEE Journal of Biomedical Health Information, 22*(3), 697706.

[6] Radoglou Grammatikis, P., Sarigiannidis, P., & Moscholios, I. (2019). Securing the internet of things: challenges, threats and solutions. *Internet Things, 5*, 41–70.

[7] Salman, T., Zolanvari, M., Erbad, A., Jain, R., & Samaka, M. (2019). Security services using blockchains: a state-of-the-art survey. *IEEE Communications Surveys and Tutorials, 21*(1), 850–880.

[8] Kumar, A., Sharma, S., Singh, A., Alwadain, A., Choi, B. J., Manual-Brenosa, J., & Goyal, N. (2021). Revolutionary strategies analysis and proposed system for future infrastructure in Internet of Things. *Sustainability, 14*(1), 71.

[9] Aloqaily, M., Boukerche, A., Bouachir, O., Khalid, F., & Jangsher, S. (2020). An energy trade framework using smart contracts: review and challenges. *IEEE Network, 11*, 1–7.

[10] Zhang, A., & Lin, X. (2018). Towards secure and privacy-preserving data sharing in e-health systems via consortium blockchain. *Journal of Medical Systems, 42*(8), 140.

[11] Rana, A., Chakraborty, C., Sharma, S., Dhawan, S., Pani, S. K., & Ashraf, I. (2022). Internet of medical things-based secure and energy-efficient framework for health care. *Big Data*, *10*(1), 18–33.

[12] Xie, S., Zheng, Z., Chen, W., Wu, J., Dai, H. N., & Imran, M. (2020). Blockchain for cloud exchange: a survey. *Comput. Electr. Eng.*, *81*, 106526.

[13] Sun, W., Cai, Z., Li, Y., Liu, F., Fang, S., & Wang, G. (Mar. 2018). Security and privacy in the medical Internet of Things: a review. *Secure Community Network*, *2018*, Art. no. 5978636

[14] Sahi, M. A., Abbas, H., Saleem, K., Yang, X., Derhab, A., Orgun, M. A., Iqbal, W., Rashid, I., & Yaseen, A. (2017). Privacy preservation in e-healthcare environments: state of the art and future directions. *IEEE Access*, *6*, 464478.

[15] Kumar, A., & Sharma, S. (2021). Internet of robotic things: design and develop the quality-of-service framework for the healthcare sector using CoAP. *IAES International Journal of Robotics and Automation*, *10*(4), 289.

[16] Yeh, K.-H. (2016). A secure IoT-based healthcare system with body sensor networks. *IEEE Access*, *4*, 1028810299.

[17] Algarni, A. (2019). A survey and classification of security and privacy research in smart healthcare systems. *IEEE Access*, *7*, 101879101894.

[18] Lwin Lwin Htay Khin Thet Wai, Nyan Phyo Aung (2019). Internet of things (IoT) based healthcare monitoring system using NodeMCU and Arduino UNO. *International Journal of Trend in Scientific Research and Development*, *3*(5):755–759.

[19] Hatzivasilis, G., Soultatos, O., Ioannidis, S., Verikoukis, C., Demetriou, G., & Tsatsoulis, C. (May 2019). Review of security and privacy for the Internet of Medical Things (IoMT). In *Proceedings of 15th International Conference Distributed Computing in Sensor Systems (DCOSS)* (p. 457464).

[20] Dhawan, S., Chakraborty, C., Frnda, J., Gupta, R., Rana, A. K., & Pani, S. K. (2021). SSII: secured and high-quality steganography using intelligent hybrid optimization algorithms for IoT. *IEEE Access*, *9*, 87563–87578.

[21] Alsubaei, F., Abuhussein, A., & Shiva, S. (Oct. 2017). Security and privacy in the Internet of medical things: taxonomy and risk assessment. In *IEEE 42nd Conference on Local Computer Network Workshops (LCN Workshops)* (p. 112120).

[22] Rana, S. K., & Rana, S. K. (2020). Blockchain based business model for digital assets management in trust less collaborative environment. *Journal of Critical Reviews*, *7*(19), 738–750.

[23] Sun, Y., & Lo, B. (Aug. 2018). An artificial neural network framework for gait-based biometrics. *IEEE Journal of Biomedical Health Information*, *23*(3), 987998.

[24] Rana, S. K., Kim, H. C., Pani, S. K., Rana, S. K., Joo, M. I., Rana, A. K., & Aich, S. (2021). Blockchain-based model to improve the performance of the next-generation digital supply chain. *Sustainability*, *13*(18), 10008.

[25] Stern, A. D., Gordon, W. J., Landman, A. B., & Kramer, D. B. (2019). Cybersecurity features of digital medical devices: an analysis of FDA product summaries. *BMJ Open*, *9*(6), Art. no. e025374.

[26] Rana, A. K., & Sharma, S. (2021). Internet of things based stable increased-throughput multi-hop protocol for link efficiency (IoT-SIMPLE) for health

monitoring using wireless body area networks. *International Journal of Sensors Wireless Communications and Control, 11*(7), 789–798.

[27] Kumar, N. M., & Mallick, P. K. (2018). Blockchain technology for security issues and challenges in IoT. *Procedia Computer Science, 132,* 1815–1823.

[28] Haghi, M., Neubert, S., Geissler, A., Fleischer, H., Stoll, N., Stoll, R., & Thurow, K. (June 2020). A flexible and pervasive IoT-based healthcare platform for physiological and environmental parameters monitoring. *IEEE Internet of Things Journal, 7*(6), 5628–5647.

[29] Baker, S. B., Xiang, W., & Atkinson, I. (2017). Internet of things for smart healthcare: technologies, challenges, and opportunities. *IEEE Access, 5,* 26521–26544.

[30] Kumar, A., Sharma, S., Goyal, N., Gupta, S. K., Kumari, S., & Kumar, S. (2022). Energy-efficient fog computing in Internet of Things based on Routing Protocol for Low-Power and Lossy Network with Contiki. *International Journal of Communication Systems, 35*(4), e5049.

[31] Jain, P., Joshi, A. M., & Mohanty, S. P. (2020). Iglu: an intelligent device for accurate non-invasive blood glucose-level monitoring in smart healthcare. *IEEE Consumer Electron Magazine, 9*(1), 35–42.

[32] Pandit, M., Gupta, D., Anand, D., Goyal, N., Aljahdali, H. M., Mansilla, A. O., & Kumar, A. (2022). Towards design and feasibility analysis of DePaaS: AI based global unified software defect prediction framework. *Applied Sciences, 12*(1), 493.

[33] Alizadeh, M., Shaker, G. Almeida, J. C. M. D. , Morita, P. P. , & Safavi-Naeini, S. (2019). Remote monitoring of human vital signs using mm-wave FMCW radar. *IEEE Access, 7,* 54958–54968.

[34] Yacchirema, D. C., Sarabia-JáCome, D., Palau, C. E., & Esteve, M. (2018). A smart system for sleep monitoring by integrating IoT with big data analytics. *IEEE Access, 6,* 35988–36001.

[35] Lilhore, U. K., Imoize, A. L., Lee, C. C., Simaiya, S., Pani, S. K., Goyal, N., & Li, C. T. (2022). Enhanced convolutional neural network model for cassava leaf disease identification and classification. *Mathematics, 10*(4), 580.

[36] Rachakonda, L., Bapatla, A. K., Mohanty, S. P., & Kougianos, E. S. (2021). Blockchain-integrated privacy-assured IoMT framework for stress management considering sleeping habits. *IEEE Transactions for Consumer Electronics, 67*(1), 20–29.

[37] Rana, A. K., & Sharma, S. (2021). Industry 4.0 manufacturing based on IoT, cloud computing, and big data: manufacturing purpose scenario. In *Advances in Communication and Computational Technology* (pp. 1109–1119). Springer: Singapore.

[38] Wang, Z. (2019). Blind batch encryption-based protocol for secure and privacy-preserving medical services in smart connected health. *IEEE Internet of Things Journal, 6*(6), 9555–9562.

[39] Rana, A. K., Krishna, R., Dhawan, S., Sharma, S., & Gupta, R. (2019, October). Review on artificial intelligence with internet of things – problems, challenges and opportunities. In *2019 2nd International Conference on Power Energy, Environment and Intelligent Control (PEEIC)* (pp. 383–387). IEEE.

[40] Huang, P., Guo, L., Li, M., & Fang, Y. (2019). Practical privacy-preserving ECG based authentication for IoT-based healthcare. *IEEE Internet of Things Journal, 6*(5), 9200–9210.

[41] Rana, A. K., & Sharma, S. (2021). Contiki cooja security solution (CCSS) with IPv6 routing protocol for low-power and lossy networks (RPL) in internet of things applications. In *Mobile Radio Communications and 5G Networks* (pp. 251–259). Springer: Singapore.

[42] Rana, S. K., Kim, H. C., Pani, S. K., Rana, S. K., Joo, M., Rana, A. K., & Aich, S. (2021). *Intelligent amalgamation of blockchain technology with industry 4.0 to improve security, book internet of things.* CRC Press, edition 1st, ISBN: 9781003140443.

[43] Boussada, R., Hamdane, B., Elhdhili, M. E., & Saidane, L. A. (2019). Privacy preserving aware data transmission for IoT-based e-health. *Computer Networks, 162,* 106866.

[44] Deebak, B. D., Al-Turjman, F., Aloqaily, M., & Alfandi, O. (2019). An authentic-based privacy preservation protocol for smart e-healthcare systems in IoT. *IEEE Access, 7,* 135632–135649.

[45] Rana, A. K., & Sharma, S. (2021). The fusion of blockchain and IoT technologies with industry 4.0. In *Intelligent Communication and Automation Systems* (pp. 275–290). USA: CRC Press.

[46] Rana, S. K., Rana, S. K., Nisar, K., Ag Ibrahim, A. A., Rana, A. K., Goyal, N., & Chawla, P. (2022). Blockchain technology and artificial intelligence based decentralized access control model to enable secure interoperability for healthcare. *Sustainability, 14*(15), 9471.

Index